MICROCOSMOS

MICROCOSMOS

JEREMY BURGESS/MICHAEL MARTEN/ROSEMARY TAYLOR

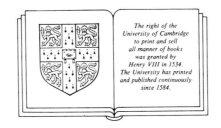

The right of the
University of Cambridge
to print and sell
all manner of books
was granted by
Henry VIII in 1534.
The University has printed
and published continuously
since 1584.

CAMBRIDGE UNIVERSITY PRESS
Cambridge
New York   New Rochelle   Melbourne   Sydney

Published by the Press Syndicate of the University of Cambridge
The Pitt Building, Trumpington Street, Cambridge CB2 1RP
32 East 57th Street, New York, NY10022, USA
10 Stamford Road, Oakleigh, Melbourne 3166, Australia

First published 1987

Printed in the United States of America

British Library cataloguing in publication data
Burgess, Jeremy
  Microcosmos.
  1. Microscope and microscopy
  I. Title   II. Marten, Michael   III. Taylor,
  Rosemary
  502'.8'2   QH205.2

Library of Congress cataloguing in publication data
Burgess, Jeremy.
  Microcosmos.
  Includes index.
  1. Microscope and microscopy.   2. Ultrastructure
(Biology)   I. Marten, Michael.   II. Taylor, Rosemary.
III. Title.
QH205.2.B87   1987   502'.8'2   87-21782
ISBN 0 521 30433 4

Text: Jeremy Burgess, Michael Marten, Rosemary Taylor,
      Mike McNamee, Rob Stepney
Editing: Michael Marten, Rosemary Taylor
Design: Richard Adams/AdCo Associates
Diagrams: Neil Hyslop

Special thanks to: Donald Claugher, Gordon Leedale, Lou Macchi,
Barry Richards, Cath Wadforth, Derek Wight

CONTENTS

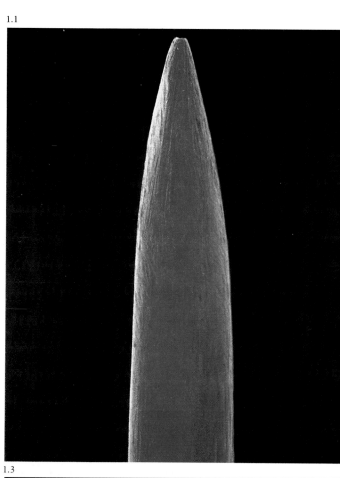

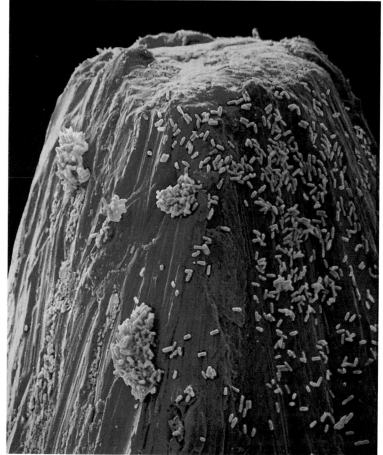

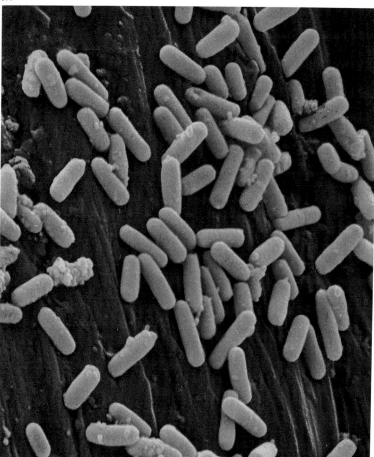

CHAPTER 1
MICROCOSMOS

IN 1683, the Dutchman Antoni van Leeuwenhoek made a startling observation in one of his regular letters to the Royal Society of London. He declared that there were more creatures living inside his mouth than there were people in the Netherlands. He was referring to the countless 'animalcules' that he had seen when examining scrapings from his teeth with one of his crude, hand-built light microscopes. Animalcules were also to be found in many other specimens – a drop of pond water, or a pinch of soil. Van Leeuwenhoek was the first person to see the teeming world of what we now know to be bacteria and protozoa.

It was not until 250 years later, in 1933, that a 26-year-old Berlin Ph.D. student called Ernst Ruska produced the first image to improve substantially on the resolving power of the light microscope. It showed a specimen of cotton fibres, and it was formed by an instrument that Ruska had been largely responsible for inventing and developing: the transmission electron microscope. Ruska's feat caused little stir at the time. For one thing, his cotton fibres were charred almost beyond recognition by the beam of electrons used to 'illuminate' them. But six years later, his brother Helmut and a colleague used a transmission electron microscope to take the picture that demonstrated the existence of viruses.

Van Leeuwenhoek's animalcules and Ernst Ruska's cotton fibres were milestones in the expansion of the visible universe. Just as telescopes have made it possible for us to see more and more of outer space, first showing us planets and then galaxies in increasing detail, so microscopes have revealed successive layers of 'inner space' – the microcosmos.

The microcosmos is extraordinarily rich and complex. The more we see of it, the more we appreciate the beauty and subtlety of nature's work. And the more we realise that events in the microcosmos can touch us all: an invisible fungus can attack a crop and cause a famine; a virus can decimate an army and change the course of history; a crack in a grain of metal can lead to the failure of a component and bring an aircraft crashing to the ground.

We pride ourselves on our visual sense, and we receive two-thirds of our sensory information through our eyes, but human eyesight is limited in its powers. The naked eye cannot, for instance, separate two dots printed less than 0.1 millimetres apart. As a result, we can just make out the blunted point of an ordinary household pin, but even the most keen-sighted person cannot begin to perceive the bacteria on the tip of the pin which might turn a trivial wound into a festering sore.

A microscope extends the power of our vision by revealing new detail, and it is able to do this because its resolution limit is smaller than that of the human eye. Most of the pictures in this book were produced with the three major types of microscope in use today. They are the light microscope, the transmission electron microscope (TEM) and the scanning electron microscope (SEM).

The light microscope has a resolution limit, in practice, of about one-thousandth of a millimetre (0.001 mm). By its use, we can see objects and organisms one hundred times smaller than the finest details visible to our unaided eyes. This means that it is possible to distinguish bacteria, but it is not possible to see an individual bacterium's shape or structure in any detail. The light microscope has undergone continuous refinement since its invention at the beginning of the 17th century; its resolving power has now reached the theoretical limit set by the wavelength of light itself. In its modern forms, it is a highly versatile instrument, and one that is simple to install and operate. It can be used to observe the gyrations of pond micro-organisms and to identify bacteria in a smear of infected blood or tissue; to spot a fault that will consign a batch of microchips to the dustbin, or to analyse a mineral sample in the evaluation of a potential oil field. It is also comparatively cheap, costing from as little as £300 up to £30 000 for the best research instruments.

The TEM has a resolution limit of one-millionth of a millimetre (0.000001 mm). This is one hundred thousand times smaller than the unaided eye can see. In practice it means that the biologist can distinguish the strands of DNA molecules inside a bacterium, while the metallurgist can observe the faults in a pin's atomic structure. This enormous resolving power carries a penalty in terms of convenience and cost. TEMs are large, static instruments costing £80 000–150 000. They require stable electricity supplies. They need to be installed in places where there are no stray magnetic fields and as little vibration as possible. They can be used to examine only extremely small whole objects – such as viruses – or ultrathin sections of larger objects, and this often means that the specimen requires lengthy and skilled preparation using expensive ancillary equipment. But by the use of TEMs we can look directly at the atomic structure of all matter, and we can study the fundamental ways in which life functions.

The SEM does not have the great resolving power of the TEM – its resolution limit is one-hundred-thousandth

1.1–1.4 The power of microscopes lies in their ability to reveal increasing detail at higher and higher magnifications. Figure 1.1 is a false colour scanning electron micrograph of an ordinary household pin, magnified ×30. The next three pictures in the series are all magnified 5 times more than the previous one, at ×150, ×750 and ×3750. They show in increasing close-up populations of rod-shaped bacteria, coloured yellow, which cling to the scored sides of the pin's blunted point. The bacteria are of the kind that might cause a scratch or pinprick to become infected.
SEMs, false colour, ×30, ×150, ×750, ×3750

1.5

1.5 Modern TEMs are designed for maximum ease of use by the operator. All the controls are to hand on the console of the instrument. The beam of electrons is generated at the top of the instrument's 'column' and travels down through the specimen to form the image. A low-power binocular light microscope can be used, as here, to monitor fine-focus adjustments to the image. Ancillary items such as the instrument's power supply and the pumps which evacuate the column are usually housed remote from the microscope itself.

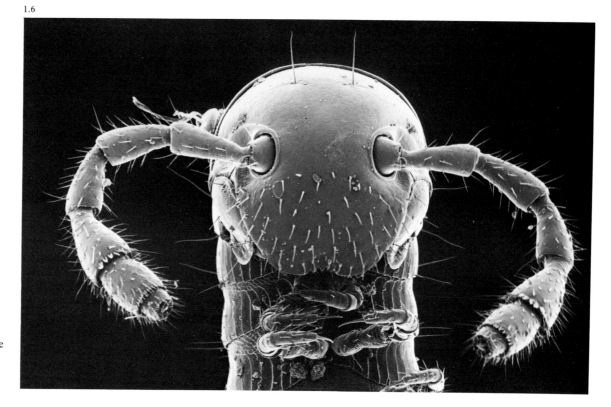

1.6 The SEM produces realistic, 'three-dimensional' images of objects and organisms. The head of this millipede is dominated by its two jointed antennae. Its body segments each carry a pair of legs ending in a single claw. The bristles decorating the head and antennae probably serve a sensory function.
SEM, ×100

of a millimetre (0.00001 mm), ten thousand times smaller than the unaided eye can see. But it possesses unique qualities which have made it a complementary instrument to the TEM. The latter is used almost exclusively to study the internal structure of things; the SEM, in contrast, shows what the external surface of an object or organism looks like. And as the pictures of bacteria on a pin illustrate, it does so with dramatic three-dimensional realism. Although the basic principles of the SEM were formulated during the 1930s, shortly after Ruska's work on the TEM, the scanning microscope had a long development and it was not until 1965 that the first commercial instrument became available. It has since become widely used in industry as well as research. This is because of its ability to look at the whole of a small object, or part of a large object, in the natural state – a microchip, an insect, or the tell-tale debris from an aircraft accident. Like the TEM, however, it is a large and expensive machine, costing £50 000–100 000.

The variety and versatility of light microscopes and the two electron microscopes have made them standard tools in many areas of science, medicine and industry. From the simple optical instrument in the school biology laboratory to the latest high-voltage electron microscope for studying the interactions of atoms, there is an instrument for most applications. The greater resolving power of the TEM and SEM has by no means rendered the light microscope obsolete. It is commonly used nowadays for routine and screening work, and in the form of the ultraviolet fluorescence microscope it plays an important role at the frontiers of biological research.

The product of every type of microscope is a magnified image of a specimen. The magnification is the ratio of the size of the image to the size of the specimen it depicts. The pictures in this book range in magnification from light micrographs of minerals magnified as little as ×7 (see Figures 7.21–26) to the electron micrograph of uranium atoms magnified ×120 million (Figure 7.2). To appreciate this enormous range, consider a familiar creature such as a butterfly. At a magnification of ×7, the average butterfly would become the size of a small bird with a wingspan of 35 centimetres (14 inches). At a magnification of ×10 000, its outstretched wings would measure 500 metres (547 yards) from tip to tip and span several city blocks. Magnified ×120 million, our humble butterfly would be metamorphosed into a beast the size of a continent, with a wingspan of 6000 kilometres (3728 miles).

As these figures imply, distances in the microcosmos are very short, and everyday units such as millimetres are of little use. Instead, two smaller units are used. The micrometre is one-thousandth of a millimetre, or one-millionth of a metre. It is about the length of the average bacterium. It is also roughly the resolution limit of the light microscope, as we have seen. Even smaller is the nanometre. It measures one-millionth of a millimetre, or one-thousandth of a micrometre. This is a very short distance indeed. Ten atoms lined up in a row would cover the span of 1 nanometre. It is also roughly the resolution limit of a good TEM.

Magnification figures can make impressive reading, but high magnification is of no use unless it is accompanied by high resolution. Any picture can be magnified indefinitely, simply by photographic enlargement. But sooner or later this ceases to reveal more detail – it produces instead a magnified blur.

Microscopes do not just magnify; because of their resolving power, they also reveal fine detail. As magnification is increased, successively finer detail is revealed until the microscope's resolution limit is reached and nothing new can be distinguished.

Many micrographs are easy to understand. This is the case with the 'realistic' pictures produced by the SEM and many of the pictures taken through light microscopes. Other micrographs are more difficult to interpret and their full appreciation requires some understanding of how the specimen has been prepared and how the magnified image of the specimen is produced.

In a light microscope, magnification is the result of light travelling from the specimen through two glass lenses, the objective lens and the eyepiece. The light may reach the objective after passing through the specimen (transmitted light), or after being reflected from its surface (incident light). Transmitted light can only be used if the specimen is naturally translucent or if it consists of a very thin section which allows light to pass through it. Most of the light micrographs in this book, particularly the biological ones, were made using transmitted light. Incident light is used when the specimen is a whole object, or naturally opaque – a microchip, for instance, or a piece of metal.

In a TEM, magnification is the result of a beam of electrons travelling down through the centre of a series of circular electromagnets, called 'electron lenses'. In one sense, the TEM resembles the light microscope – the electron beam is transmitted through the specimen. But the electrons must travel in a vacuum, and the method by which the image is formed is not analogous to that which occurs in a light microscope. Electrons are not very penetrating particles, and TEM specimens must therefore be very thin indeed to enable the electron beam to pass through them. The typical TEM specimen is about 1/1000th of the thickness of a single page of this book. This requirement determines the main characteristic of transmission electron micrographs: they represent only a tiny section of a whole object.

The SEM can be likened to an optical microscope used in incident light mode. A very fine beam of electrons, again travelling in a vacuum, is focused by electron lenses onto the specimen. As the beam strikes it, other electrons are emitted from the specimen's surface and radiate outwards. These 'secondary electrons' are collected by a detector and used to produce a point of light on a television screen. The image is built up by scanning the electron beam over the specimen in a series of lines and frames, called a raster scan. As the electron beam moves over the specimen, it produces a series of points of light on the television screen, so that a complete image is built up. The magnification is a direct result of the ratio of the specimen area scanned to the area of the television screen. If the electron beam is set to scan the whole of a microchip, the magnification might be of the order of ×50; if it scans just a few of the microchip's multitude of circuits, the magnification might be ×10 000.

Because the secondary electrons are emitted from its surface layers, the SEM specimen does not need to be a thin section. It can be a whole object, up to a size determined by the microscope's specimen chamber. Specialist SEMs have been built for the examination of objects as large as a whole knee joint and even a gun barrel. Secondary electrons are not the only particles produced when the primary beam strikes the specimen surface. X-ray photons are also emitted, and so occasionally are photons of visible light. And some of the primary electrons may be 'back-scattered', or reflected, off the specimen. These radiations can also be used to form an image. X-ray imaging is particularly useful, because the X-rays are emitted with energies characteristic of the elements that make up the specimen's surface. It is therefore possible, for example, to map the positions of the different elements in an alloy or composite material (see Figure 8.40).

Before it becomes a specimen for microscopy, an object must usually be prepared in some way. For the SEM this preparation may amount to no more than coating the object with an extremely thin layer of gold to improve its surface conductivity and emission of secondary electrons. Biological SEM specimens may also need a simple freezing step to stabilise them against the vacuum inside the instrument. Many light microscope and TEM specimens, on the other hand, are thin sections and require elaborate and skilled preparation.

Sections make it possible to view opaque objects and examine their internal structure. And in the case of both the light microscope and the TEM, the thinner the section, the higher the potential resolution of the image. Methods have therefore been devised for slicing or grinding objects of all kinds until they are wafer-thin. In the case of light microscope specimens, such sections are usually about 5 micrometres (1/200th of a millimetre) thick. TEM specimens need to be about 200 times thinner at 100 nanometres (1/10 000th of a millimetre) or less.

1.7

1.7 In order to produce a clear image, specimens for the SEM are usually coated with a metal such as gold to increase their ability to reflect electrons and their electrical conductivity. The photograph shows a microchip being covered with a gold layer only 5–10 nanometres thick by a process called sputter-coating. The specimen is in a vacuum chamber containing argon gas at low pressure. A flat plate of pure gold is positioned a few centimetres above the microchip, and a voltage of about 1000 volts is applied between it and the gold plate. As a result, the gas ionises, and argon ions strike the gold plate so forcefully that gold atoms are ionised and dislodged from it. These gold ions then stream downwards and form a uniform layer on the surface of the specimen. The ion plasma is seen in the picture as the bright glow; its shape is controlled by a circular magnet around the gold plate at the top of the picture.

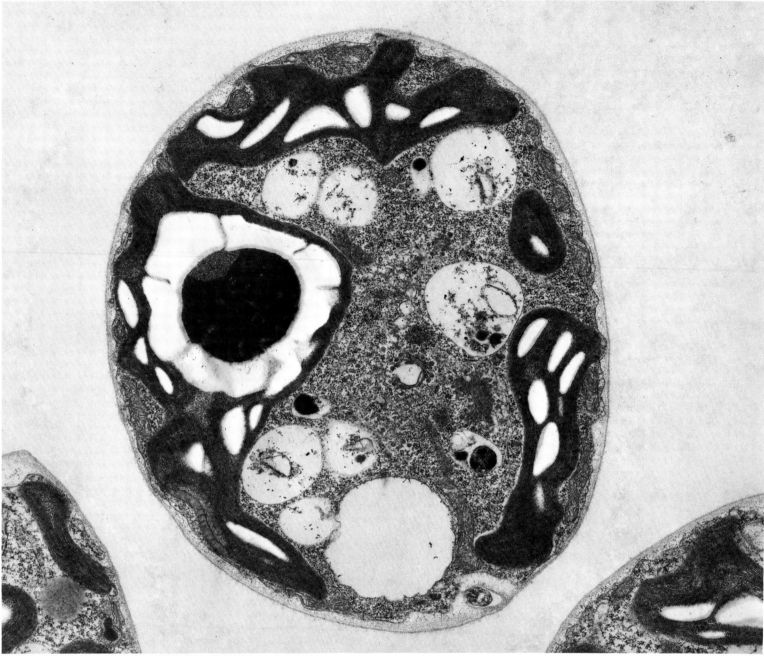

1.8 The problems of image interpretation are at their most acute in the case of thin sections photographed using a TEM. This picture typifies such micrographs. It is a section of the unicellular green alga, *Chlamydomonas asymmetrica*, at ×20 000 magnification. In looking at this image, as with all sections, it is important to remember that what appears in the picture is a two-dimensional representation of a tiny part of a three-dimensional object. The cell is about 10 micrometres in diameter; the section represented by this picture is only 70 nanometres thick, or rather less than 1 per cent of the thickness of the whole cell.

Because the section is two-dimensional, the cell appears round although it is spherical, and the pure white areas, which are sections of ellipsoidal starch grains, appear elliptical. Furthermore, not all the components of the cell are visible in this tiny sampling of it; there is no nucleus in the picture, for example, although the cell would certainly contain one. The dark areas around the periphery of the cell are sections through chloroplasts. There are five distinct dark areas visible, although two nearly touch. It is possible that all these areas are in fact part of a single chloroplast which is sufficiently

undulating to disappear from view in such a thin section. Likewise, the circular grey regions in the cytoplasm, the vacuoles, may indeed be discrete spherical spaces; on the other hand, some of them at least may coalesce in a plane which is not included in this section. Microscopists habitually examine large numbers of sections to establish three-dimensional relationships when this is important. Equally, they habitually select sections for photography which are as 'perfect' as possible, and fairly describe the 'typical' structure of their object. This picture is free from technical defects such as staining dirt, scratch lines from

a faulty microtome knife, and folds. It is a good 'typical' picture of *C. asymmetrica* insofar as it clearly shows the pyrenoid in the chloroplast (the black area within the white ring). This was why it was taken. It is a poor picture of an algal cell, because it does not show the nucleus; however, it is highly probable that the dozen or so sections which would give a view of the nucleus would in fact not contain the pyrenoid in this particular cell. In this sense, micrographs are created by their photographers; they do not happen on their own.

TEM, stained section, ×20 000

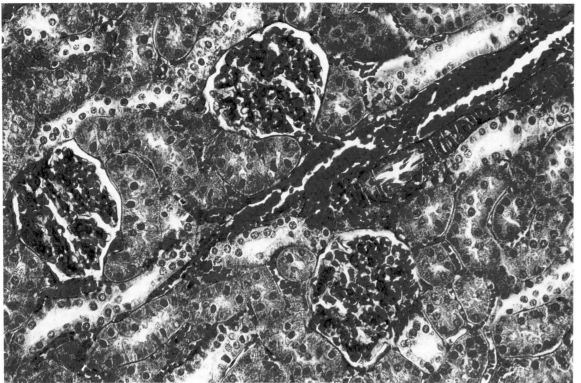

1.9 The thin section of human kidney tissue in this light micrograph contains a blood vessel, which runs diagonally from top right towards bottom left, and three glomeruli. Our kidneys have thousands of glomeruli, each of which is a tiny bundle of tightly packed capillaries which filters the blood of excess water and poisonous substances. The bright colouring of the tissue results from the stain used during the preparation of the specimen.
LM, trichrome stain, ×940

Biological thin sections are usually transparent and so low in contrast that they need to be stained in order to produce a satisfactory image. Different stains can emphasise particular features in a specimen. For example, staining a section of a cell with uranium acetate, a heavy metal salt, makes the genetic material – the DNA and RNA – appear black in a transmission electron micrograph. In light microscopy of human and animal tissue, a dye called eosin is widely used and colours the cytoplasm of cells various shades of pink.

Since colour is a property of light, the image produced by a light microscope is seen in colour, although it may be photographed in black-and-white. The colours themselves may be due to staining, or to the form of illumination chosen by the microscopist, or they may be the natural colours of the specimen. Natural coloration is common in pictures of small living organisms, but even here the operator can decide whether the creature is seen against a black background (dark field illumination) or a light one (bright field illumination). And in the former case, filters can be used to tint the background any chosen colour (Rheinberg illumination).

Other illumination techniques change the colour of a specimen completely. The use of polarised light, for example, produces colours by the process of birefringence. The technique can produce very colourful images of crystalline materials, including common chemicals like vitamin C or aspirin, and this has made it popular with many amateur microscopists. It is also used quantitatively. The colours displayed by a section of rock of known thickness, for example, result from fundamental properties of the minerals from which the rock is composed. A geologist can therefore use the technique as a precise tool for identifying rock and mineral types.

Electron micrographs are always originally monochromatic. Unlike photons of visible light, electrons do not 'carry' colour and therefore can tell us nothing about a specimen's coloration. But electron micrographs can be artificially coloured by computer, photographic, or hand-tinting techniques. Such 'false colour' can make it easier to distinguish particular structures in a specimen, but it is usually added for aesthetic effect and it need bear little or no relation to the natural colour of the object or organism portrayed.

The 300 or so pictures in this book represent only a tiny sample of modern microscopy. We have not attempted to include examples of every technique or every specialist microscope; nor have we tried to cover every application of microscopy. The aim has been to produce a book of some of the best and most informative micrographs, and through them to introduce the microcosmos.

(The figure captions in this book all end with a 'technical line' in small type. This gives information about the type of microscope used to produce the image, and the magnification. Also included where relevant and known are details about the illumination technique, the type of section and the stain. All scanning electron micrographs in the book have been produced by secondary electron imaging except where otherwise stated. Abbreviations used in the technical lines are: LM – light microscope; SEM – scanning electron microscope; TEM – transmission electron microscope; STEM – scanning transmission electron microscope; HREM – high-resolution electron microscope; DIC – differential interference contrast; H & E – haematoxylin and eosin.)

CHAPTER 2
THE BODY

Over the past three centuries, microscopy and medicine have advanced hand in hand. They continue to do so. Because structure and function are intimately linked, progress in our understanding of body processes in health and disease has to a considerable extent depended on our ability to see increasingly fine structural detail in our organs and tissues. And it is advances in our understanding rather than new equipment or drugs that underlie the most important developments in medicine.

The 16th century brought the first systematic dissection of animal corpses, and growing interest in the *post mortem* examination of the human body. From what was readily visible with the unaided eye, physicians began to relate disease to the gross appearance of the internal organs. In the 17th century microscopists started to probe the hidden structure of biological materials. When Leeuwenhoek died in 1723, he bequeathed 26 microscopes. With each microscope came an example of the specimens for which it had been built: among them were blood, sperm, hair and muscle.

With an increasing use of microscopes in the late 18th century came an interest in texture, and the realisation that organs are composed of various forms of 'tissue' – a word that first came into use at this stage. The basic material of which we are composed was thought to be fibre, at times dense, and at others loosely woven.

In the mid-19th century, the greater resolving power of microscopes revealed the cell as the basic building block of tissue and the site of the disordered processes underlying disease. Progress was also dependent on advances in the preparation of specimens, particularly in the cutting of thin sections, and in the increasing use of dyes to reveal structure by staining. In the 1870s and 1880s, almost every coloured substance known was tried as a stain in microscopy; and the advent of synthetic dyes greatly added to the limited range of plant extracts available.

By the end of the 19th century microscopists had realised that a dark-staining nucleus was found in almost all cells, and that it was perpetuated when cells divided. They had seen ribbon-like chromosomes within it, and by 1900 had directly observed the process of fertilisation. Only with these observations of the fundamental features of cells did it become clear how male and female make equal contributions to their offspring; and how the characteristics of one generation are transmitted in an orderly way to the next.

More recently, the advent of electron microscopy has revolutionised our concept of the cell, revealing complicated structures within it that account for many of its specialised functions. It was not until transmission electron microscopy of muscle cells in the 1950s, for example, that we learned how muscle contraction works.

But medical microscopy is above all a working, practical technology. With most cancers, for example, symptoms and gross appearance are only an uncertain guide. Examination by microscope is still the way malignancies are confirmed. And surgeons often hang fire in the operating theatre until the potentially abnormal specimens they have taken can be examined. This is the province of histology – the study of normal tissue – and of histopathology – the study of its diseased counterpart.

The process of preparing a histological specimen for light microscopy begins by taking tissue representative of the feature under study, and fixing it (often in formalin) to prevent putrefaction and degeneration. The sample is then placed in increasing concentrations of alcohol to remove all water. Once dehydration is complete, the alcohol is removed by immersion in an organic solvent such as xylene. The next stage is to embed the tissue in a medium that provides strength and support. In light microscopy, it is usual to impregnate the specimen with paraffin wax. The solidified block of tissue is then cut by a microtome. This machine shaves off a series of sections, each around 4–5 micrometres thick, forming a ribbon of samples. The ribbon is floated in water, and sections picked up on a glass microscope slide, ready to be dried and stained.

The use of stains is crucial in histology: the history of the science is one of improved staining techniques as much as of improved microscopes. Haematoxylin and eosin are stains in routine use. All specimens are first treated using these two dyes. Their effect is to stain the nuclei of cells blue and the cytoplasm shades of pink. A battery of dozens of stains can then be brought to bear to pick out individual structures.

Given the immense diversity of tissues within the body, and the range of microscopical techniques available to study them, the images in this chapter cannot provide a comprehensive coverage. And although microscopy is an essential tool in pathology, the body as it appears in these pages is predominantly a healthy one. The pictures are also mostly of human specimens. However, where structure and function are essentially the same, images of tissue from other animals – usually other mammals – are included.

2.1 This light micrograph shows a wafer-thin section through one of the folds of the wall of the jejunum – the second part of the small intestine. The area covered is about half a millimetre across. The intestine at this point is deeply folded and lined with many thousands of finger-like projections, or villi, which protrude into the hollow tube along which food passes. Around 20 villi can be seen in this picture. Villi are all of much the same length, but the unusual angle of this section makes them appear of differing size. Each villus is composed of a mucous membrane, here stained pink, and a central core of blood vessels and small lymphatic channels. The mucous membrane is made up largely of columnar cells, called enterocytes. It is just possible to distinguish the individual columnar cells aligned in a single row to form the surface of each villus. The stain used in the preparation of the specimen has dyed the nucleus at the base of each cell a dark red. Enterocytes are involved in the breakdown of food, but their major role is in the absorption of its constituent nutrients. Scattered among the enterocytes are goblet cells, which have stained pale blue. Goblet cells secrete the mucus lining that protects the intestine from self-digestion. Nutrients absorbed by the enterocytes pass into the rich network of blood capillaries and lymphatic vessels at the centre of each villus. Extending along the base of the villi is the *muscularis mucosae*, containing blue-stained smooth muscle cells.
LM, trichrome stain, ×430

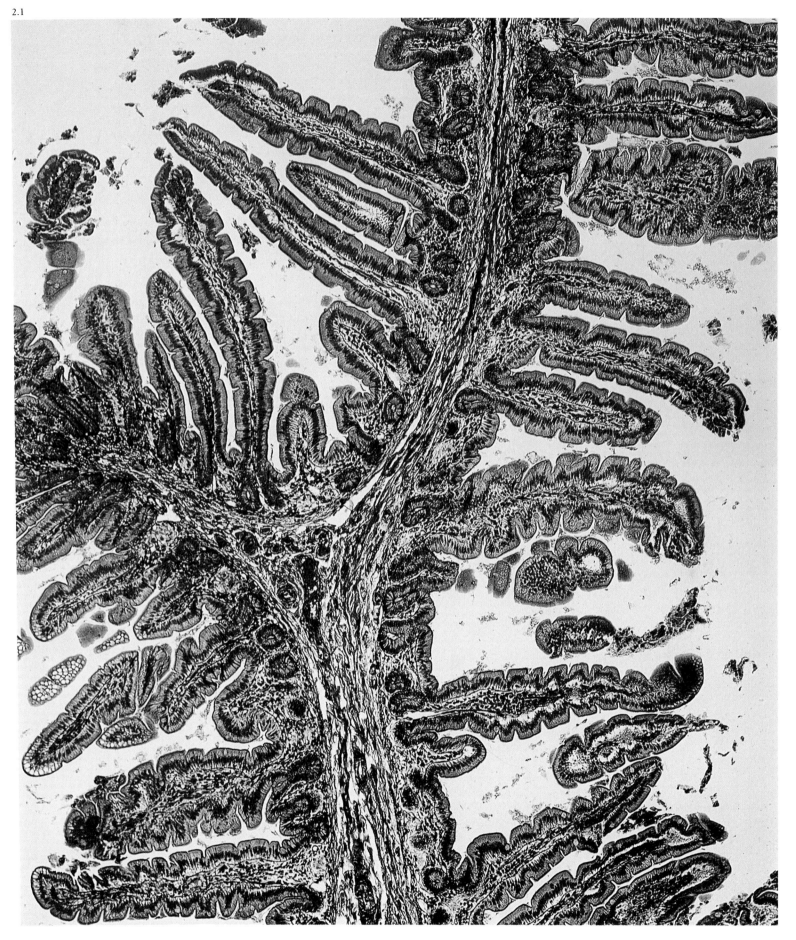

REPRODUCTION

Sexual reproduction involves the fusion of two specialised cells which, when combined, provide the genetic blueprint for the development of a unique individual. The bringing together of male and female sex cells, or gametes, is the result of years of preparation, and a formidable piece of plumbing. The photographs on the following pages show three crucial parts of the process: the development of sperm, the development of eggs, and the moment of fertilisation.

The generation of sperm in the tubules of the testes is a lengthy but continuous and limitless process. Each ejaculation contains roughly 500 million sperm.

In contrast, a woman's few hundred thousand potential eggs, or oocytes, are all present in her twin ovaries at birth. This relative scarcity dictates their far less prodigal use: a ration of one per menstrual cycle. And – to make the best of limited opportunities – the slow two-week ripening of the chosen egg, its release into the oviduct, and its eventual reception in the uterus (if fertilised) are accompanied by hormonal orchestration of exquisite timing.

2.2 Sperm are produced in the hundreds of coiled seminiferous tubules packed into each testis. This scanning electron micrograph shows a transverse section of a single tubule. Towards its circular rim, relatively large, round and undifferentiated cells – called spermatogonia – can be seen. Certain of these cells develop over roughly two months to become sperm, complete with tails, which form the tangled mass in the hollow middle of the tubule. This process involves a stage of double cell division (meiosis), which reduces the number of chromosomes each sperm contains to half that found in the body's other cells. Other spermatogonia undergo normal cell division (mitosis), so maintaining the tubule's population of potential sex cells.
SEM, ×720

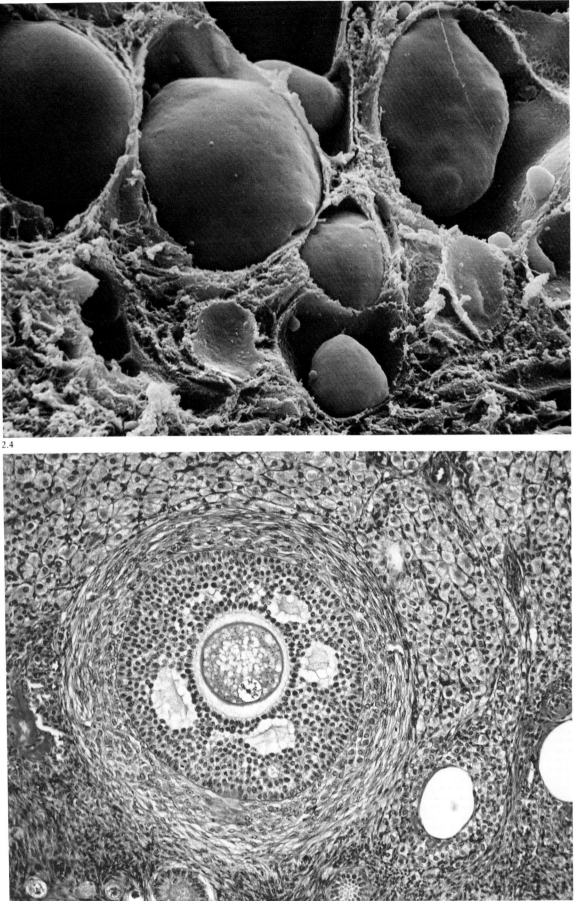

2.3 The outer portion of the ovary – the cortex – contains sacs (follicles) which house eggs at various stages of development, from the most rudimentary oocyte to the single mature cell released at ovulation. In this scanning electron micrograph, follicles nestle in the human ovary, separated by connective tissue. Each month, about 20 primitive follicles, along with the eggs inside them, start the process of maturation. But only one follicle develops to the point where a mature egg is shed. The remainder simply degenerate.
SEM, ×1615

2.4 This light micrograph is of a cross-section through a single follicle in the ovarian cortex. The small, perfectly rounded oocyte at centre has reached its final size, but is not yet fully mature. A thin blue circle – the *zona pellucida* – surrounds it. Then comes a granular layer of cells, 10–12 deep at this stage, in which irregular sacs containing follicular fluid can be seen. These sacs eventually coalesce into a single large cavity, the follicular antrum. More elongated connective tissue cells form a halo around the follicle. These cells have already begun to secrete the hormone oestrogen, which leads to a thickening of the lining of the uterus, ready for the implantation of a fertilised egg. At ovulation, the mature follicle bursts open and the egg is expelled, only to be enveloped by finger-like projections that channel it into the oviduct, where it lingers for a few days in anticipation of the arrival of sperm. The crater formed by the ruptured follicle continues to play a part. Its cells are reorganised to form a temporary endocrine gland that secretes the hormone progesterone, which further prepares the uterus for implantation.
LM, ×110

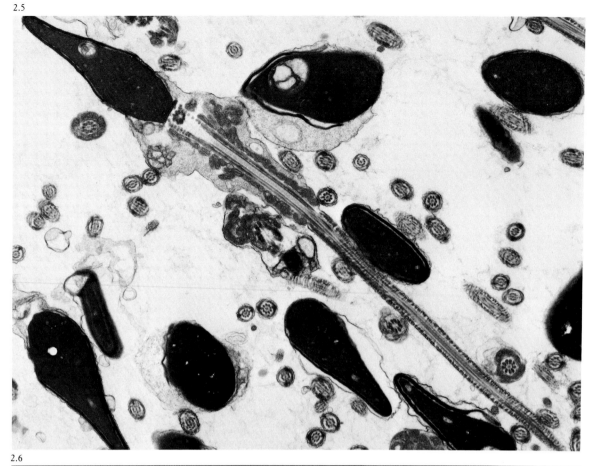

2.5 The staining technique used to prepare this thin section of human sperm causes their heads to appear black. One sperm is seen complete with its tail, running from top left to bottom right. It is about 65 micrometres long. The arrow-like nucleus containing the 23 chromosomes of the male sex cell is tipped with enzymes that aid penetration of the egg. The translucent material around the sperm's 'neck' consists of remnants of the cytoplasm of the spermatogonia from which the sperm has developed. Next comes the flagellum, the whip-like tail whose thrashing enables the sperm to propel itself. Running the length of the tail, in its centre, are nine coarse strands, wrapped around with fibrous ribs to provide strength. Mitochondria, cell components which generate the energy to power the sperm, can be seen tightly packed against the core of the tail in the section (forming roughly one third of its total length) nearest the head.
TEM, negative stain, ×1300

2.6 Since its eggs are fertilised in water, and therefore easily observed under the microscope, the sea urchin is a classic subject of research. Despite the wide evolutionary gap, the process of fertilisation in humans and sea urchins is similar, both biologically and in appearance. In this light micrograph, the cell membrane of a sea urchin egg has been penetrated by a sperm, which entered six minutes before the picture was taken. The entry point can be seen just above the three o'clock position. It used to be thought that the sperm was the active agent, migrating through the egg's cytoplasm to fuse with the central nucleus. But this picture shows that the normally circular nucleus has become elongated as it moves to meet the sperm.
LM, DIC, magnification unknown

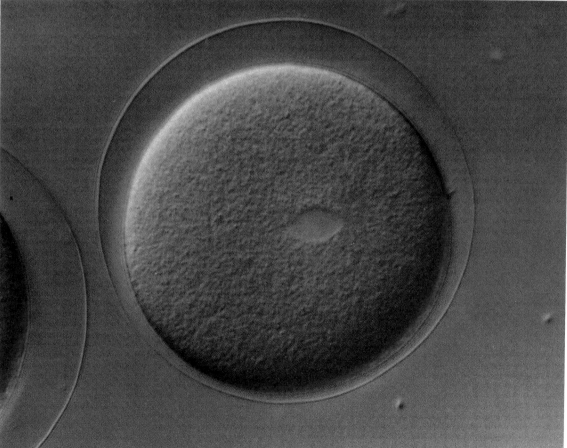

2.6

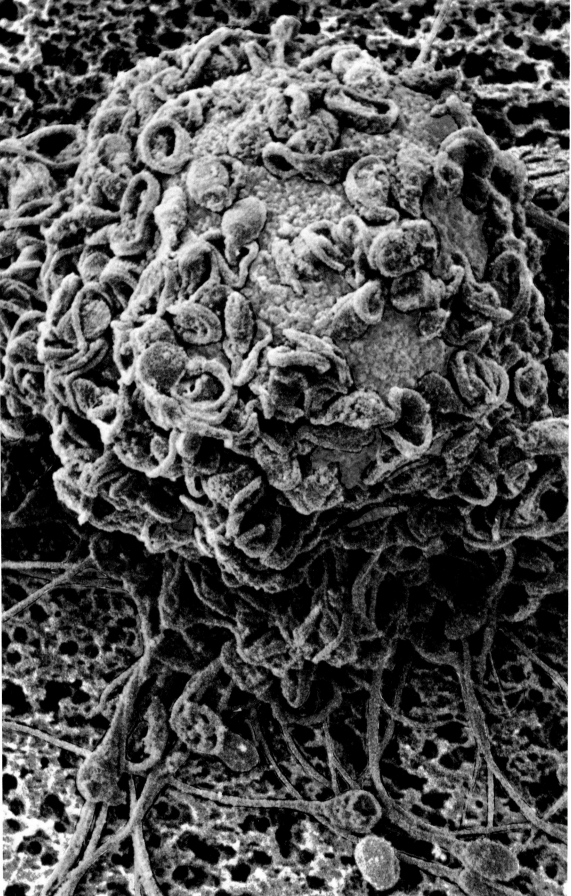

2.7 A human egg (pale pink) is almost overwhelmed as sperm scrummage to be the first and only one to penetrate it. As soon as one sperm has succeeded, a rapid chemical change causes the outer membrane of the egg to thicken, so obstructing the entry of other sperm. The tadpole-like structure of sperm is clearly seen in this scanning electron micrograph. So too is the enormous size difference between male and female sex cells: at roughly one tenth of a millimetre in diameter – a speck just about visible to the naked eye – an egg is 200 times the size of a sperm. Identical twins develop from the splitting of a single fertilised egg; and non-identical twins from two separate eggs fertilised at the same time. The process of twinning has nothing to do with the entry of more than one sperm into the same egg.

SEM, false colour, ×2650

VISION

Our awareness of the world
requires receptor cells that
respond to changes in the
environment, a means of
transmitting this raw information
to the brain, and processes that
analyse and impose meaning on
what would otherwise be just a
'booming, buzzing confusion'.
Though there is far more to
perception than the sense organs,
they are clearly fundamental; and
in humans the eyes are the most
fundamental of all.

Light enters the main chamber
of the eye through the pupil,
which varies in aperture from the
size of a pinpoint to that of a pea.
The light is then focused by the
lens onto the retina, at the back of
the eyeball, where it bleaches the
pigment of detector cells. This
chemical change generates
patterns of electrical activity which
are transmitted to the brain. The
three scanning electron
micrographs on these pages show
structures forming part of the
optics of the eye, and the nerve
cells underlying visual sensation.

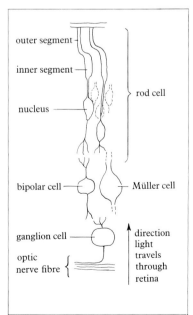

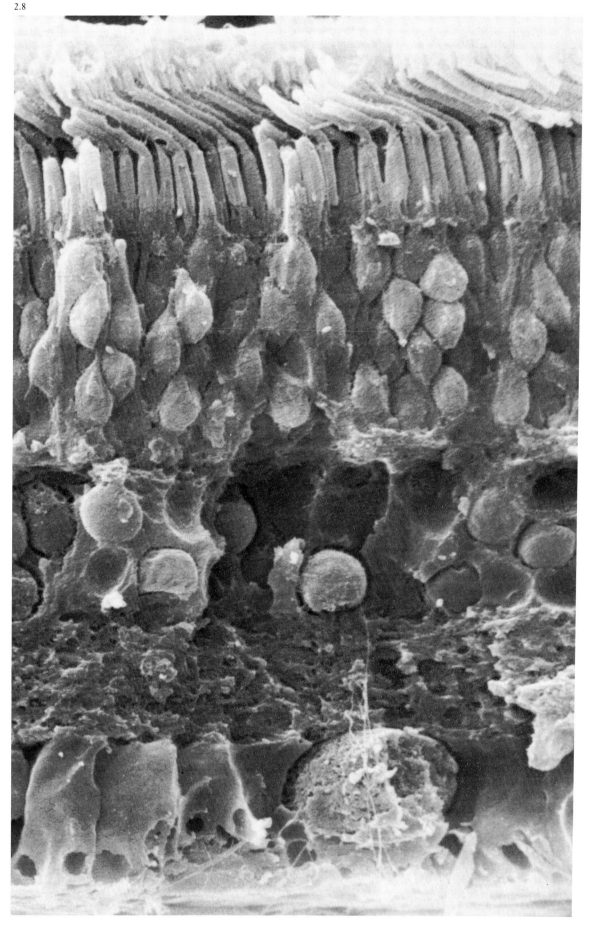

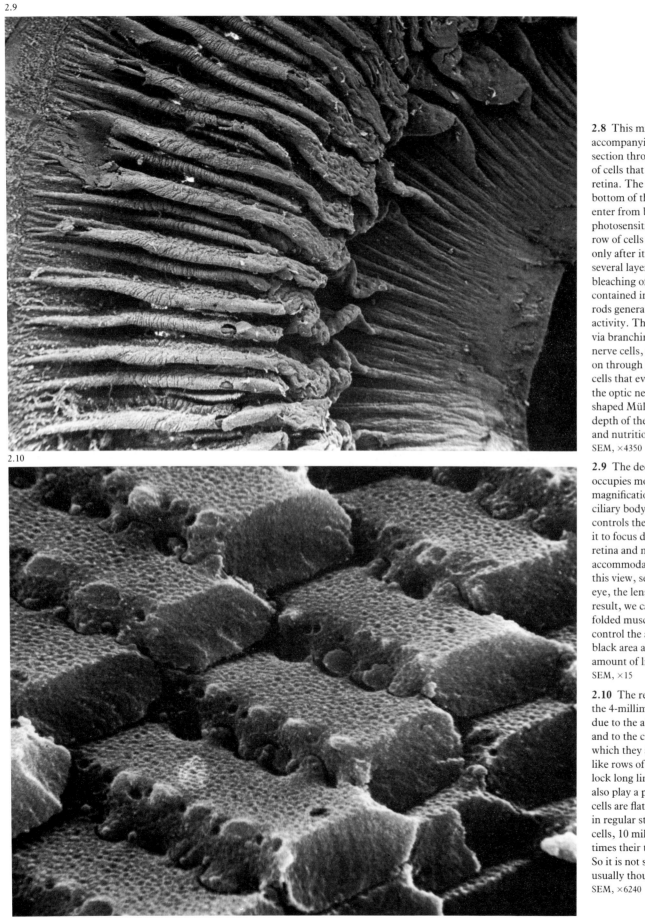

2.9

2.10

2.8 This micrograph and accompanying diagram show a cross-section through the complex sandwich of cells that makes up the mammalian retina. The front of the retina is at the bottom of the picture, so light would enter from below and strike the photosensitive rod receptors (the neat row of cells at the top of the image) only after it has first passed through several layers of cell bodies. The bleaching of light-sensitive pigment contained in the outer segments of the rods generates patterns of electrical activity. These signals are transmitted via branching nerve endings to bipolar nerve cells, which pass the information on through junctions with ganglion cells that eventually form the fibres of the optic nerve. Large, irregularly-shaped Müller cells extend through the depth of the retina, providing support and nutrition.
SEM, ×4350

2.9 The deeply-folded tissue that occupies most of this low-magnification image is called the ciliary body. It normally encircles and controls the lens of the eye, flattening it to focus distant objects onto the retina and making it more spherical to accommodate those that are closer. In this view, seen from the back of the eye, the lens has been removed. As a result, we can see through to the less-folded muscles of the iris, which control the aperture of the pupil (the black area at bottom right) and so the amount of light that reaches the retina.
SEM, ×15

2.10 The remarkable transparency of the 4-millimetre thick lens of the eye is due to the absence of nuclei in its cells, and to the crystalline precision with which they are arranged. The zipper-like rows of ball-and-socket joints that lock long lines of cells together may also play a part. In cross-section, lens cells are flattened hexagons, arranged in regular stacks. The length of lens cells, 10 millimetres, is about 2000 times their thickness, 5 *micro*metres. So it is not surprising that they are usually thought of as fibres.
SEM, ×6240

HEARING

Our organ of hearing – the cochlea
– is structured like a snail's shell.
It is a tapering, fluid-filled spiral
cavity 3.5 centimetres long.
Membranes running the length of
the spiral divide it into three
parallel ducts, the upper and lower
ducts being connected at the apex.

2.11 In this light micrograph, the coils
of the cochlea are seen in section.
The surrounding bone appears blue
and the fluid-filled chambers of the
spiral are white. The two membranes
are clearly seen. The upper membrane
appears tissue-thin; the lower – the
basilar membrane – is more
substantial. The three ducts they form
are also easily distinguished.

Positioned in the middle duct – on
the basilar membrane that forms its
floor – is the organ of Corti. Here, in
cross-section, it appears as a small
wedge shape in each of the cochlea's
coils. Rather like a spiral piano
keyboard, the organ of Corti runs the
full length of the middle duct. It is this
part of the cochlea that actually
responds to sound. At its top is a shelf-
like projection – the tectorial
membrane – and below it rows of
sensory cells rooted in the basilar
membrane. These 'hair cells' – seen in
more detail in the scanning electron
micrographs on the opposite page – are
ultimately responsible for our
perception of pitch and intensity.

Sound waves striking the ear drum
are transmitted via three small bones
(the ossicles) to an opening, called the
oval window, at the start of the upper
duct of the cochlea. There the
vibration is translated into movement
in the cochlear fluid. This movement
causes the basilar and tectorial
membranes to rise and fall, producing
shearing forces that stimulate electrical
activity in the hair cells lying between
them. Fibres from the hair cells
bundle together to form the auditory
nerve, which leads to the brain.

The loudness of a perceived sound
seems to depend on the *magnitude* of
vibration in the basilar membrane.
Intense noises produce a large
displacement, and generate a fast rate
of impulses from the hair cells.

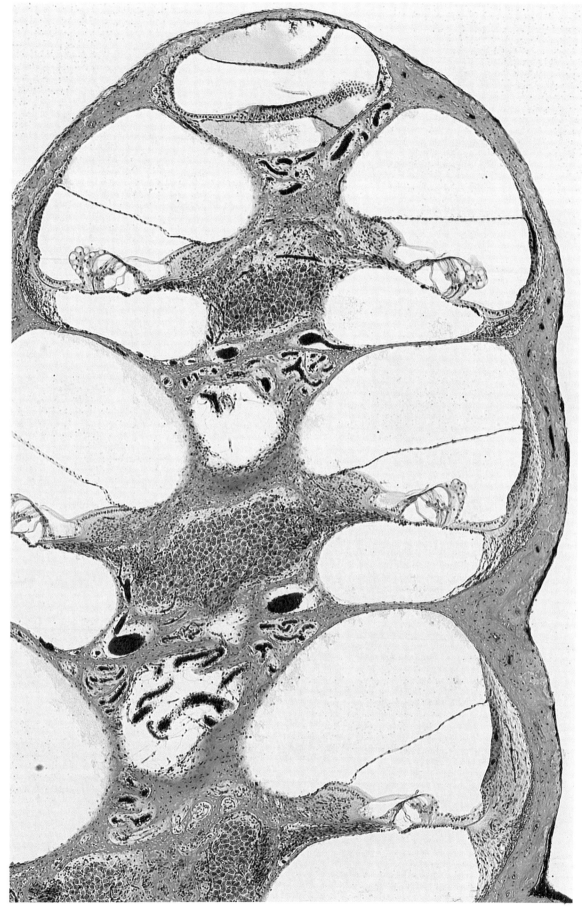

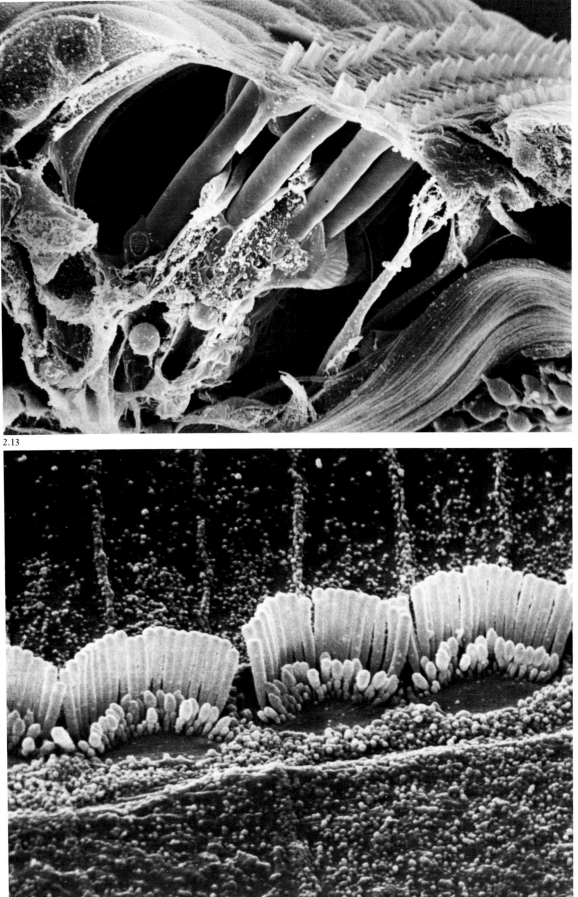

2.12

2.13

Perception of pitch is more complicated, but seems to involve the *position* along the basilar membrane at which cell firing is greatest. We know that hair cells in different regions of the cochlea are sensitive to different frequencies. High-frequency sounds have a short wavelength and generate a movement in the cochlear fluid which peaks near the broad origin of the spiral tube, exciting maximally the hair cells at that point. With lower tones, the wave amplitude takes longer to build up, reaching its peak near the tapering end of the tube, and producing the greatest stimulation among a different population of cells. Apparently, the organ of Corti's greatest sensitivity corresponds to the frequency of the human scream.
LM, trichrome stain, ×234

2.12 More than 20 000 hair cells, each having as many as 100 separate hairs, make up the organ of Corti. Their task is to translate mechanical movements caused by their displacement into electrical impulses. Towards the top of this micrograph are four rows of hairs, three of them stoutly supported by the pillar-like Deiter cells below. The basilar membrane has become buckled during preparation of the specimen and appears as a wavy sheet at bottom right. Normally, the tectorial membrane would lie above the hairs, making contact with the tallest of them.
SEM, ×1970.

2.13 At greater magnification, the hair cells are seen to produce several neatly ordered rows of hairs, which are also called stereocilia. The tall hairs are clearly arranged according to height. The smooth area just in front of the smaller stereocilia is also part of the hair cell surface.
SEM, ×7585

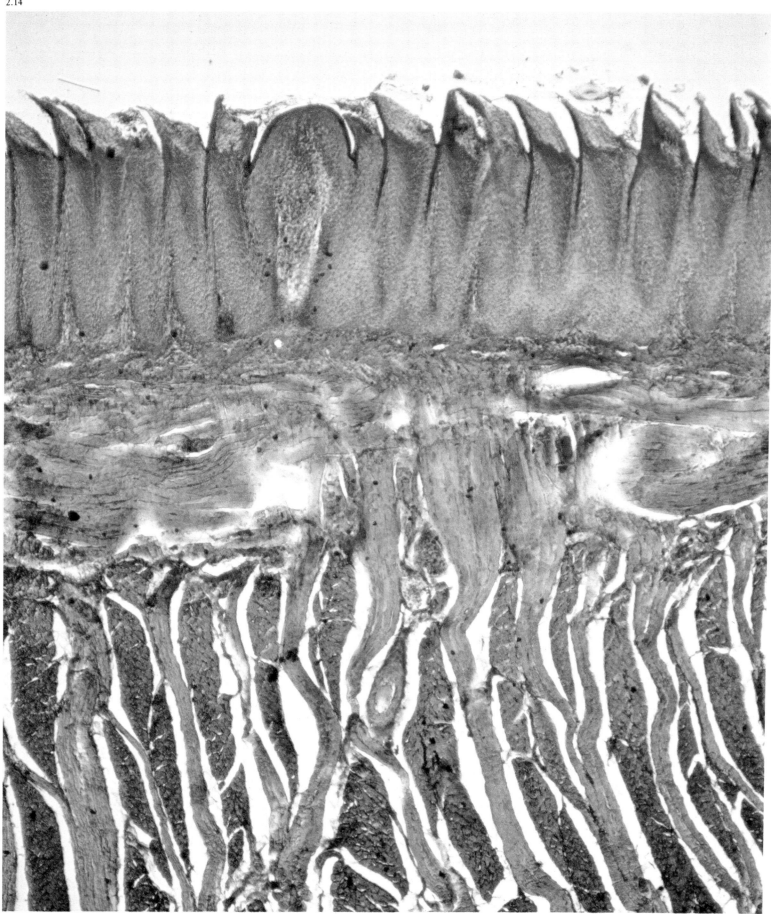

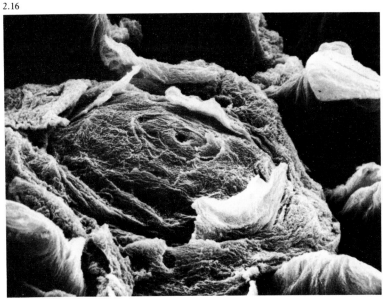

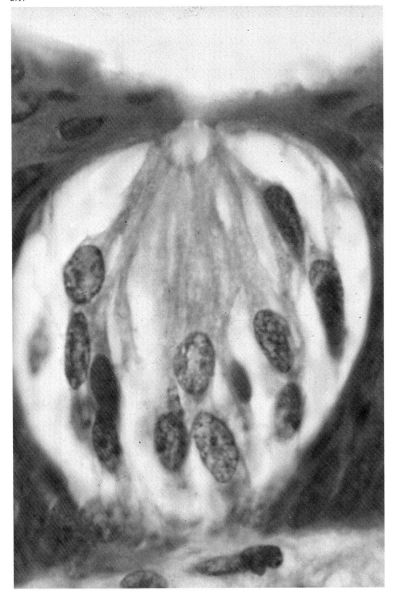

TASTE

Much of the tongue is made up of
muscle, which gives it the
extensive repertoire of movements
we require for speech and for the
manipulation of food. But its
mucosal surface is highly
specialised for sensory perception,
and for the appreciation of taste in
particular. The tongue is covered
in numerous tiny projections,
called papillae. These projections
differ slightly in shape, size and
colour. Filiform papillae are thin,
whitish strands. Fungiform
papillae are larger and have a good
blood supply, which makes them
appear red. They are large enough
to appear as small red dots when
one looks at one's tongue in the
mirror. Larger still are the
circumvallate papillae – so-called
because they are surrounded by a
deep cleft, or trench. There are
eight to ten circumvallate papillae
at the back of our tongues and it is
here that most of our taste buds
are found. Four different elements
of taste have been identified:
sweet, bitter, acid and salt. Many
of the more sophisticated
discriminations we make probably
derive more from our sense of
smell than of taste.

2.14 Tough muscle fibres make up the
lower two thirds of this section of
a rabbit's tongue. In the top third is
the uneven surface composed of many
feathery, filiform papillae, which have
stained pink. Contained within the
larger, rather bulbous protrusion is a
taste bud – well attuned, one might
imagine, to the delights of lettuce.
LM, ×85

2.15 In humans, taste buds line the
deep cleft which surrounds
circumvallate papillae. This higher
magnification section through a human
tongue reveals six taste buds, two
above and four below. Collections of
glands at the base of the cleft (far
right) secrete a fluid that dissolves food
and so aids perception of its taste.
LM, H & E stain, ×580

2.16 This scanning electron

micrograph shows a single fungiform
papilla. The small crater at its centre is
a 'taste pore', which gives access to the
pit of a taste bud.
SEM, ×1100

2.17 This section through a taste bud
shows up to 30 spindle-shaped cells
stretched across it from top to bottom.
Nerve connections suggest that only
certain of these cells are taste
receptors; others may play a
supporting role. Despite recognition of
the four basic elements of taste, which
seem to be located in different parts of
the tongue, it has so far not been
possible to distinguish different types
of taste bud.
LM, H & E stain, ×1680

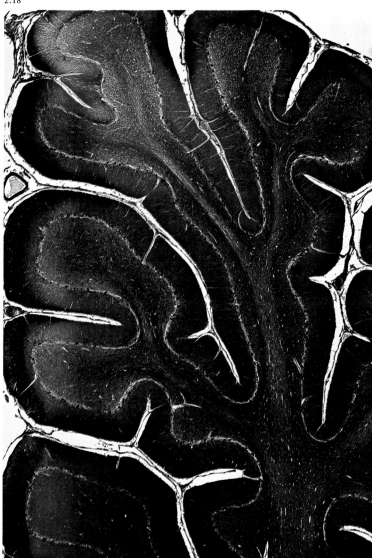

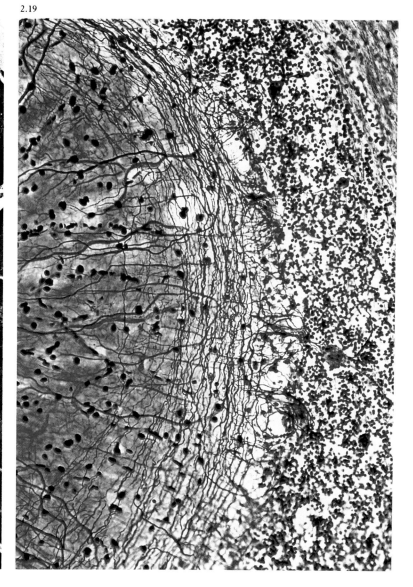

NERVOUS SYSTEM

From a reflex knee jerk to playing the violin, from the urge to eat lunch to understanding quantum physics, the nervous system's communication network of 100 000 million cells extends throughout the body and mediates every thought and action. All nerve cells, or neurones, consist of a cell body, containing the nucleus, and a range of extensions, the cell processes. From the cell body project numerous thin 'arms', each ending in branching 'fingers', or dendrites, which reach out to contact the cell bodies of other neurones. There is also usually one far longer projection: the axon. In the case of neurones that serve the periphery of the body, the axon may extend unbroken for more than a metre.

The nervous system is conventionally divided into two parts: the central nervous system (CNS), made up of the brain and spinal cord, and the peripheral nerves. The CNS is a concentrated mass of interconnected nerve cells which integrate information, analyse it and make decisions. The peripheral nervous system, on the other hand, consists mostly of nerve processes – the axons and dendrites – that link the CNS with the body's sensory receptors and its effector organs, notably muscles and glands. The receptors detect information about the external and internal environment; and effector organs produce a coordinated response according to CNS directives.

2.18 The human cerebellum coordinates fine movement, posture and balance. It is a large, convoluted structure at the base and back of the brain. In this low-magnification light micrograph, each fold of brain tissue is seen to consist of layers – cells in the outer portion staining more darkly. Unusually-shaped neurones called Purkinje cells inhabit the whiter band between the inner and outer layers. Densely packed nerve fibres form the central core of each fold.
LM, Glee's Marsland silver stain, ×60

2.19 The infinite, tangled complexity of nerve pathways is revealed by this higher-magnification view of the cerebellum. Within the human brain, there may be as many as 50 000 dendrites making contact with a single cell body. Each is capable of conveying a message from another neurone. Some of these connections make it more likely that a cell will fire; others are inhibitory. Whether or not a particular nerve cell fires depends on the net effect of the excitatory and inhibitory messages. Neurones in the cerebellum are responsible for our manual dexterity.
LM, silver stain Bodian, magnification unknown

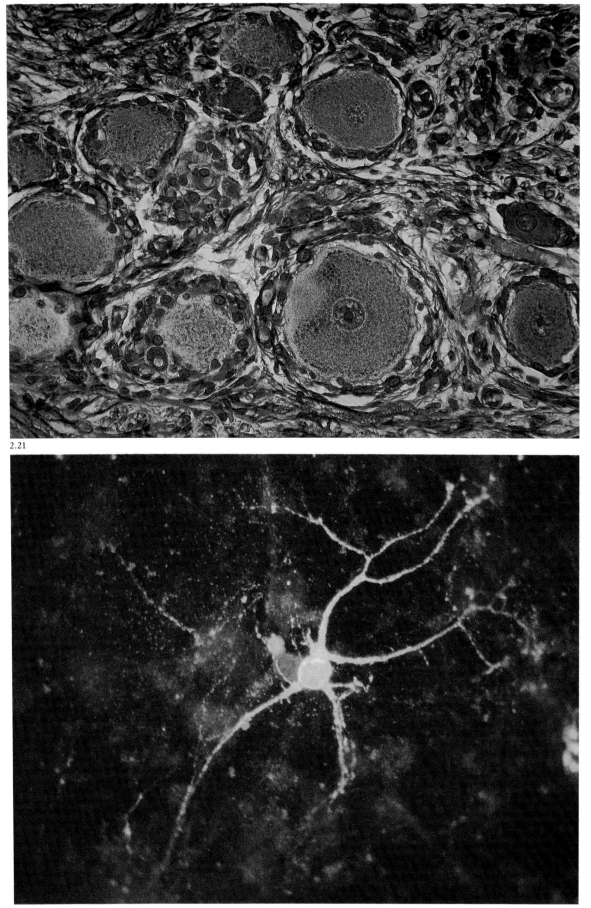

2.20 Tightly-packed neuronal cell bodies appear as large, pink circles in this light micrograph of a human spinal ganglion – a collection of nerve cells found just outside the spinal cord at the point where it is joined by a spinal nerve. Centrally placed within each of the largest cell bodies can be seen a smaller circle, the cell's nucleus. Surrounding each cell body is a single layer of flattened, supporting 'satellite' cells, stained dark pink. The simple knee-jerk reflex is one example of a movement involving a spinal ganglion. The message that the knee has been tapped passes from sense organs in the skin along the axon of a sensory neurone, through the spinal ganglion which houses the cell body of that neurone, and into the spinal cord. There, the sensory neurone is linked via an intermediate nerve cell to a motor neurone which runs out of the spinal cord and back to the knee with the signal that a muscle must contract.
LM, trichrome stain, ×710

2.21 The discovery that the body produces its own opiate-like chemicals marked an important breakthrough in our understanding of addiction, pain perception and pain relief. Immunofluorescence microscopy of the kind illustrated here played a vital part in tracing the nerve systems in which these substances operate. In this technique, a fluorescent dye is used to label an antibody molecule that will attach itself to the substance under study. A piece of tissue is then exposed to the antibody. When the specimen is viewed under ultraviolet light, the location of the target substance is revealed by fluorescence. Here, yellow-green fluorescence shows the presence of the opiate-like chemical enkephalin in the cell body and cell processes of a single neurone. The neurone is an example of a *multipolar* nerve cell, which has many dendrites projecting from the cell body.
LM, fluorescence, ×110

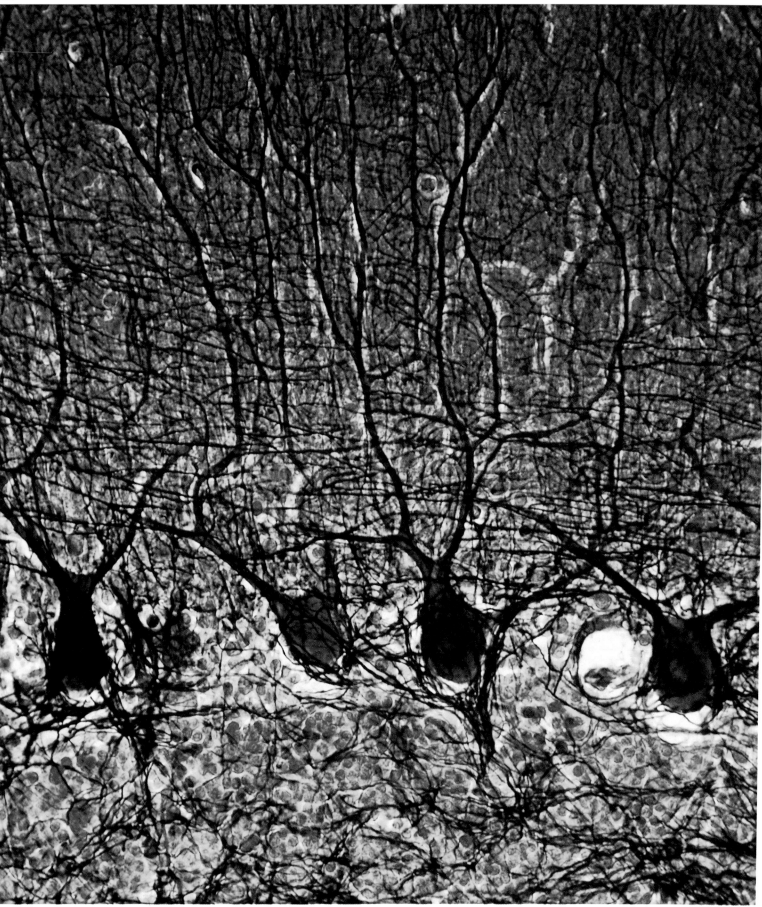

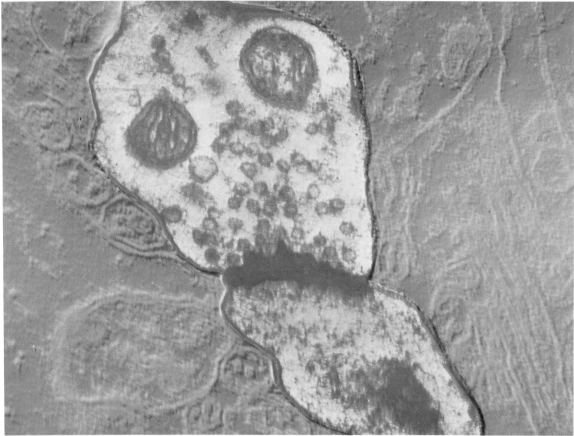

2.22 The outer zone of the cerebellum forms the upper two-thirds of this micrograph. Below it are four spider-like Purkinje cells, which are among the largest neurones in the body. The cerebellum receives input from areas of the brain responsible for initiating movement, and from the body's sense receptors, but we do not know precisely how its cells integrate motor and sensory information and coordinate fine movement.
LM, silver stain ×980

2.23 This electron micrograph shows a junction, or synapse, between two neurones of the human cerebral cortex. Although the knob-like ending of the upper nerve seems to touch the post-synaptic membrane of the one below, there is a minute gap between them, which is coloured red. Transmission of information along nerves is by electrical impulse. But its passage from one nerve to another involves neurotransmitter chemicals. The small red and orange specks in the upper nerve ending are sac-like vesicles containing neurotransmitter. When activated, the biochemicals are released to diffuse across the synaptic cleft. They then interact with receptors on the next neurone, triggering or inhibiting a new electrical impulse. The two larger circles towards the top of the image are mitochondria – organelles that produce energy for the cell.
TEM, false colour, ×78 300

2.24–2.25 A motor neurone activates muscle cells through end-plates – small swellings that form the terminals of the axon. These neuromuscular junctions are portrayed in a light micrograph (Figure 2.24) and by scanning electron microscope (Figure 2.25). Where fine motor movement is required, one neurone may control one muscle fibre. But for cruder control, as here, a single axon can innervate several muscle fibres. The passage of instructions from nerve to muscle relies on the same chemical processes as synaptic nerve–nerve transmission.
2.24 LM, gold chloride impregnation stain, ×170
2.25 SEM, magnification unknown

2.24

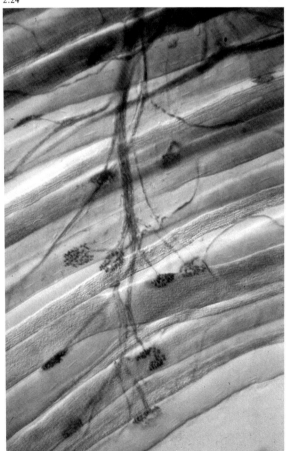

2.25

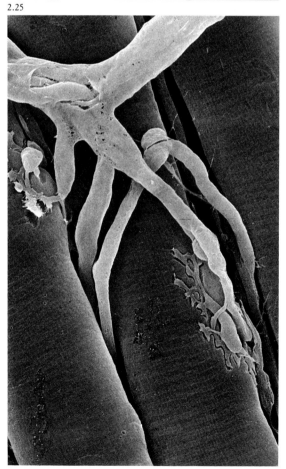

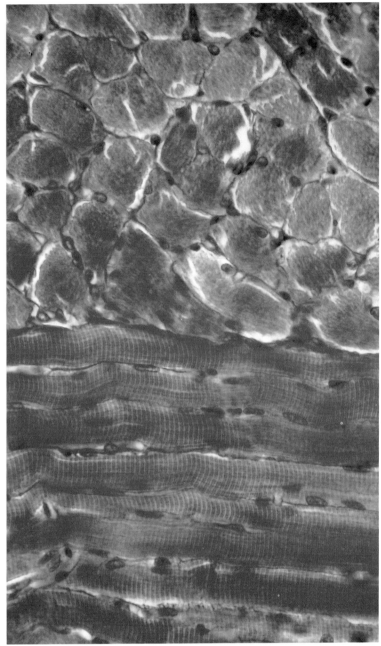

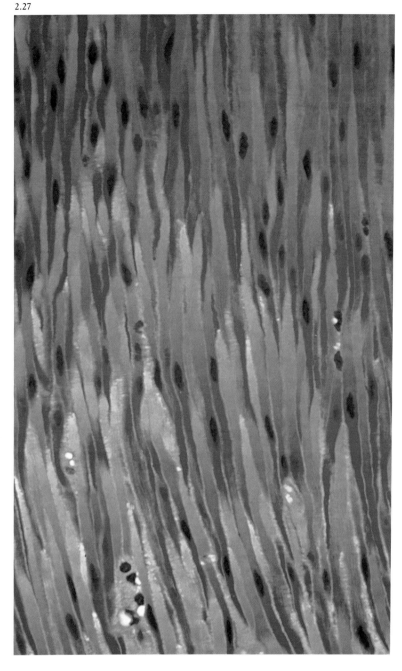

MUSCLE

Muscle cells form three distinct kinds of tissue. The most familiar, because it moves our limbs, is skeletal muscle. It is also called striated muscle, because of its prominent banding. This banding is absent in so-called smooth muscle, which performs largely automatic processes such as the contraction of blood vessels and of the gut. Similar to skeletal muscle in appearance, but unique in its capabilities, is cardiac muscle, which is capable of setting up its own rhythm of contraction, independent of nervous system control. It is unique too in its tirelessness – contracting every second for a lifetime.

2.26 Human striated muscle is seen here both in cross-section (top) and longitudinal section (bottom). Skeletal muscle cells are long – hence their description as muscle *fibres*. Each fibre has several nuclei spaced at intervals along it. Both parts of this micrograph show that these dark-staining nuclei are positioned at the edge of the cell. Running the length of each fibre, barely visible in the longitudinal part of the section, are contractile proteins arranged in thin threads called myofibrils. More prominent is the banding *across* the fibres, caused by myofibrils being composed of alternating light and dark sections.
LM, ×620

2.27 The spindle-shaped cells that form smooth muscle are grouped in irregular bundles. They do not generally have to contract with the same force as skeletal muscle, nor with the same speed or precision. There is only one nucleus per cell.
LM, H & E stain, ×445

2.28 The small black dots in this cardiac muscle cell are granules of animal starch, or glycogen. Glycogen is chemically broken down to glucose, and glucose in turn yields energy to power muscle contraction. Most of this chemistry takes place in mitochondria – the elliptical structures at left and bottom right. The picture also shows clearly the long threads of the myofibrils and their alternating light and dark sections.
TEM, stained section, magnification unknown

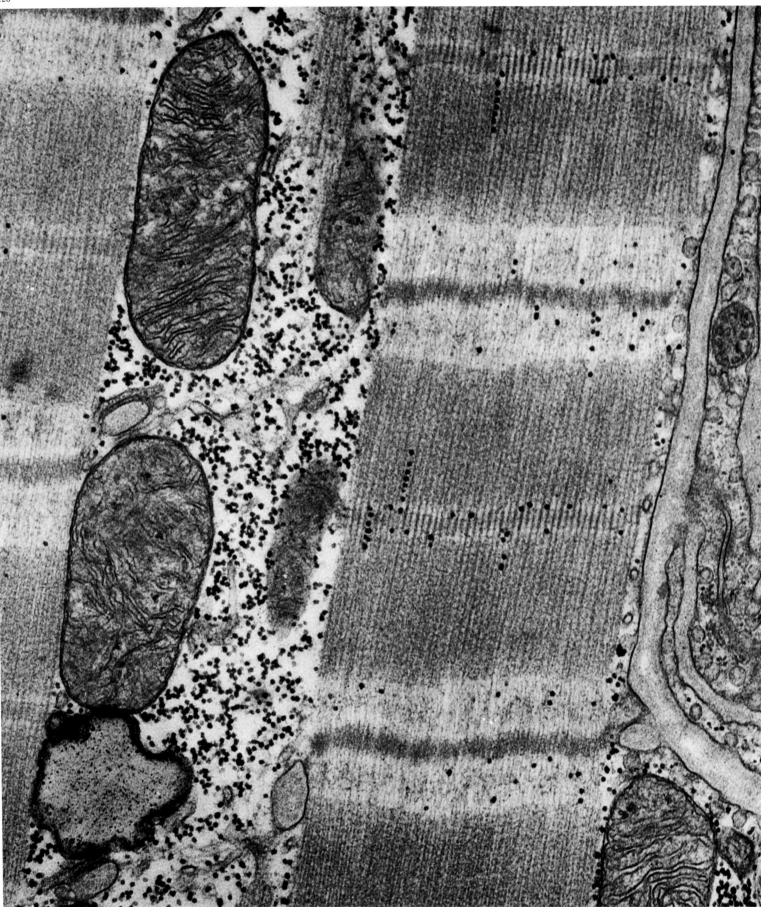

BONE

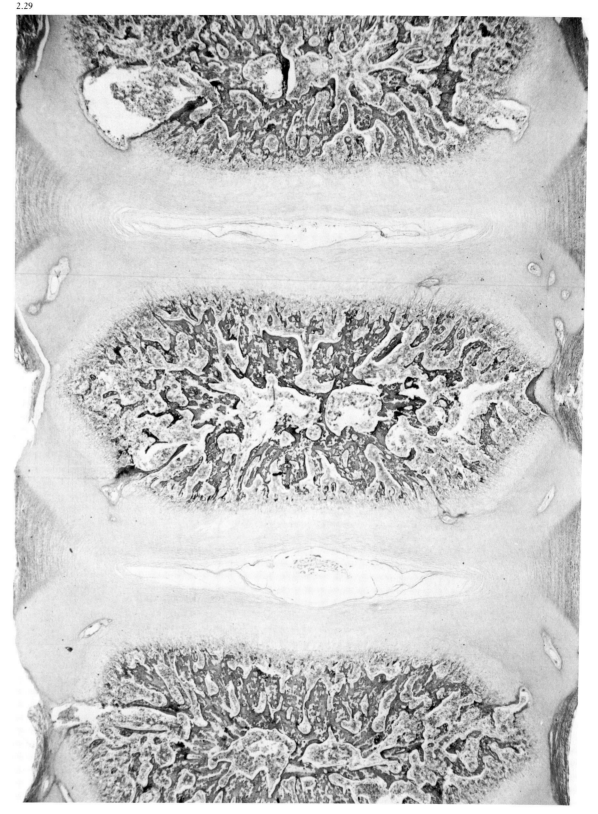

The human skeleton gives the body shape and support. It is made up of 206 rigid bones, together with the cartilage (gristle) that lines the joints and forms the cushion-like discs separating the vertebrae. The whole structure is supremely well engineered. In typical long bones, such as the femur, compact bone forms the solid outer wall of the shaft, while the interior is composed of more spongy material. This gives a high resistance to mechanical stress for a low overall weight. Bones also anticipated efficient architectural structures such as the dome (the skull), column (the femur, humerus) and arch (the foot).

As well as providing form, the skeleton plays a crucial role in maintaining the body's internal biochemical environment – acting as a store of minerals for the body to call on in times of need. In their hollow centres, bones also house marrow, which acts as the production line for both white and red blood cells.

The hardness of bone derives from its high content of calcium salts. But bone is far from inert, for all its mineral strength. Depending on the demands of the body, it is constantly growing or being resorbed. The calcium is laid down in a matrix called osteoid, which is secreted by 'bone-building' cells, or osteoblasts. These cells encircle canals that carry blood vessels.

2.29 In this low-magnification light micrograph of the human backbone, the tissue in the centre of each vertebra appears sponge-like in section. Between the three areas of bone are two intervertebral discs – in effect hydraulic shock absorbers that protect the spine and brain from unnecessary jolting. The discs are made mostly of cartilage strengthened with collagen fibres, which are flexible but extremely strong and do not stretch. These fibrocartilage discs enable the vertebrae to be connected to each other in a way that allows some degree of movement, without any sacrifice in strength. The fine concentric layers of cartilage, which appear light grey, surround a white space – the *nucleus pulposus* – filled with a thick fluid. It is this unusual form of liquid connective tissue that acts as the hydraulic element in the shock absorber. Discs do not actually slip out from between the bones of the spinal column; but with advancing age there is a tendency for the ring of cartilage to weaken, allowing the nucleus pulposus to be extruded. This produces the painful condition known as a slipped disc. LM, ×18

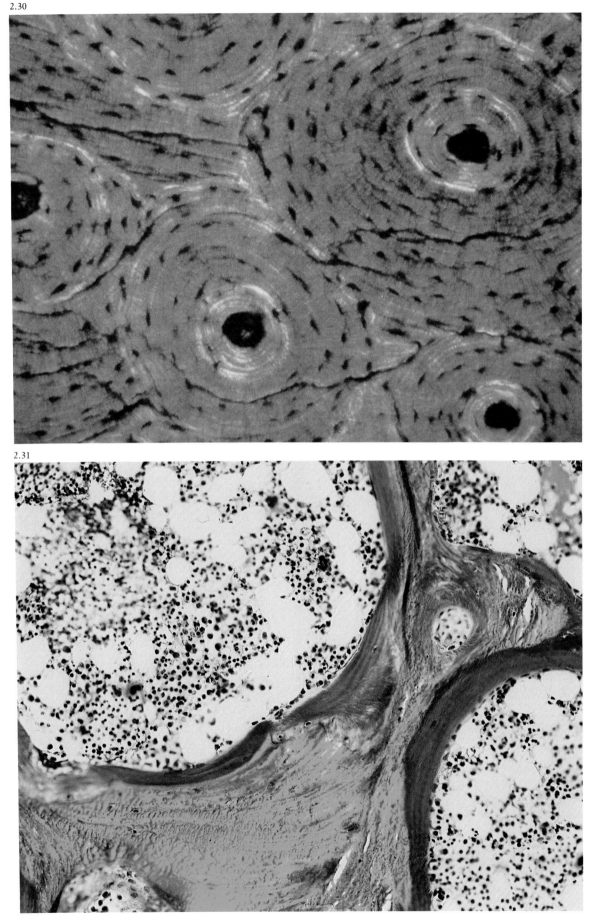

2.30

2.31

2.30 The shaft of long bones such as the femur is composed of compact bone, seen in cross-section in this micrograph. Concentric circles of bony material surround large channels (black) that contain blood vessels, lymph vessels and nerves. These channels, called Haversian canals, run the length of our bones. Looking like lines of stitching around each canal are lacunae – literally holes in the bone. Each one is occupied by an osteocyte (an osteoblast cell that has become embedded within the bone matrix). Each osteocyte sends out fine projections to make contact with its neighbours, giving the cells a spider-like appearance. Haversian canals start out wide, and gradually become filled in as layers of bone are formed. Each canal, together with its surrounding bony plates, forms a Haversian system. Haversian systems follow the lines of stress within a bone, acting like scaffolding poles. Living bone is constantly being remodelled. Large osteo*clast* cells break it down, and its components are removed. As this happens, old Haversian systems disappear and new ones come into being.
LM, Schmorl's picro thionin stain, ×210

2.31 The appropriate use of stains can reveal osteomalacia, or 'softening of the bones', a condition known as rickets when it occurs in children. If uncorrected, it leads to abnormal bending of the bones and deformity in adult life. Osteomalacia is caused by a deficit of vitamin D, usually because of poor diet and inadequate exposure of the skin to sunlight. It also occurs in clinical conditions such as chronic kidney failure. The problem is that osteoblast cells continue to secrete their organic matrix, osteoid, but deposition of calcium crystals is delayed. There is therefore accumulation of unmineralised osteoid tissue. In this micrograph, active osteoblasts appear as the tiny dark granules, and fully mineralised bone is green. Unmineralised osteoid appears as the brown layer in between.
LM, undecalcified resin section, Goldner's trichrome stain, magnification unknown

RESPIRATION

The respiratory system exists to enable oxygen from the atmosphere to be absorbed by the blood, and to allow carbon dioxide to be excreted in return. Once absorbed, oxygen is conveyed to the tissues, where it is needed so that carbohydrate fuel can be burned to provide energy. The respiratory tract is essentially a system of branching tubes, made largely of cartilage, which convey air from the mouth and nose to the sites in the lungs where gas exchange takes place. Although our lungs can contain 5 litres of air when fully expanded, each breath typically holds 0.5 litres.

The respiratory tract proper begins with a single tube, the trachea, which then forks to form the right and left main bronchi. Each bronchus in turn divides into two or three smaller tubes, and then into innumerable finer airways called bronchioles. Because exchange of gases requires a large surface area, bronchioles end in clusters of tiny sacs, or alveoli. The presence of so many cavities gives a healthy lung the appearance of an air-filled sponge.

The body's tube-like organs – such as the gut and the respiratory tract – are lined with tissue rather like the skin that covers our exterior. This lining, called epithelium, consists of layers of closely-packed cells with little cementing substance in between. Beneath the epithelium is a supporting layer of connective tissue, usually containing glands. The epithelium and this layer form the mucosa.

2.32 Clumps of fine hairs, called cilia, protrude from the tops of specialised epithelial cells lining the bronchial passages. Along with the cilia, this scanning electron micrograph shows numerous 'goblet cells'. Their function is to release mucus onto the surface of the epithelium. Rhythmic movement of the cilia serves to move bacteria and other particles away from the gas-exchanging parts of the lung and towards the throat where they can be swallowed or coughed up.
SEM, ×950

2.33 In contrast to the healthy state in the previous image, this micrograph shows a cancer of the bronchus – the most common form of lung cancer. Such cancers are caused by smoking and are almost always fatal. They frequently occur near the point where the trachea forks to form the two bronchi, since this area suffers from heavy deposition of the carcinogenic tars contained in tobacco smoke. The disorganised region of malignant cells at bottom right is invading the normal ciliated epithelium at left and top. Cancers consist of primitive cells which have not developed any specialised function. In this instance, for example, no cilia are present. The prime characteristic of cancer cells is their uncontrolled growth.
SEM, ×950

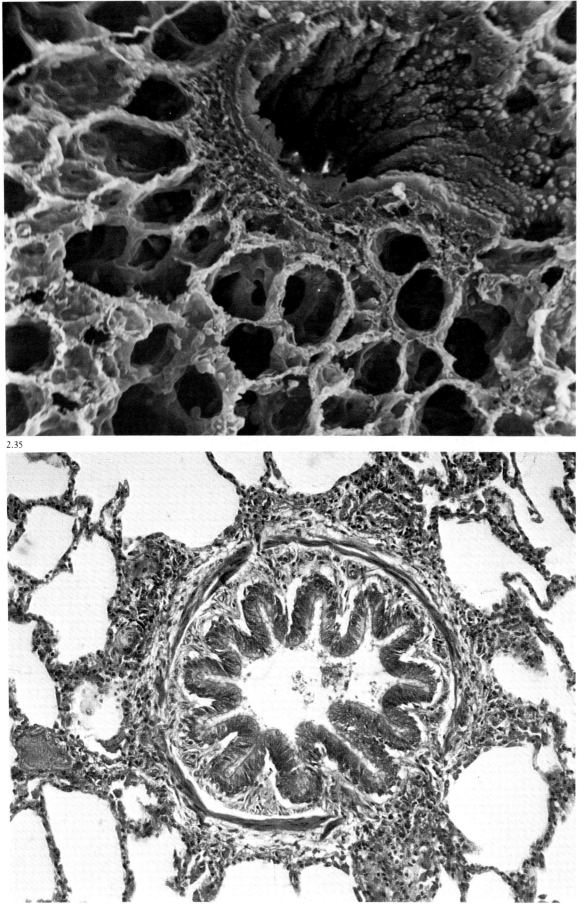

2.34 The fine lacework walls of the alveoli, and a single bronchiole (top), are seen in this scanning electron micrograph. The human lungs contain around 700 million alveoli, giving a total surface area the size of a tennis court. The walls of the alveoli are extremely thin, usually being composed of just a single layer of cells. Immediately next to them (although not visible in this specimen) are fine blood capillaries whose walls consist of a single layer of cells. This means that the barrier between air on the one side and blood on the other is only 0.3 micrometres thick; oxygen and carbon dioxide readily diffuse across. As the conducting airways reach into the depths of the lungs, they become increasingly narrow; and their walls are composed more of smooth muscle and less of cartilage. The inner surface of the bronchiole shown here appears corrugated. The structure of its highly folded epithelial lining is seen more clearly in the light micrograph below. SEM, ×280

2.35 This micrograph shows almost the same view as the previous scanning electron micrograph, but here the bronchiole is seen in cross-section, with the white areas of the alveoli around it. At 0.5–1.0 millimetre in diameter, bronchioles are the smallest of the lungs' airways. The deeply folded surface of the bronchiolar epithelium is clear. It is made up of simple rows of column-like cells, surrounded by a ring of smooth muscle cells which controls the bronchiole's diameter. Abnormal contraction of bronchiolar smooth muscle, as in asthma, severely restricts the flow of air and gives rise to symptoms of breathlessness and wheezing. Beyond the encircling smooth muscle are the alveoli. Their thin walls are composed of flattened epithelial cells and more rounded cells which are thought to secrete surfactant, a detergent-like material that lowers surface tension and improves the elasticity of the lungs. LM, H & E stain, ×180

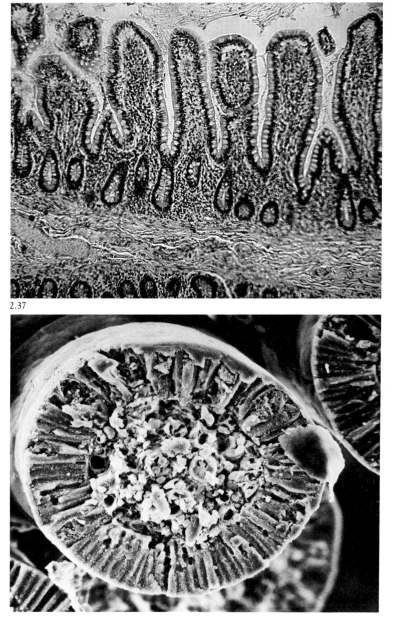

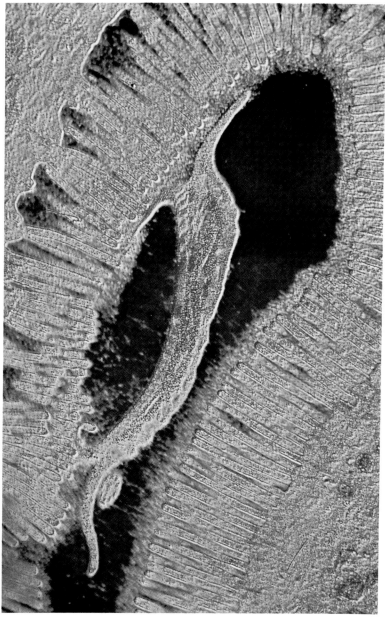

2.37

DIGESTION

Before food can be used by an organism, it must first be broken down. The gastrointestinal tract, or gut, is the site of both the digestion and absorption of food. Digestion is achieved primarily by means of enzymes, proteins which catalyse specific chemical reactions. It occurs mainly in the stomach and the upper part of the small intestine, the duodenum. Enzymes are secreted into the

intestinal contents by glands in the gut wall, and by separate organs, such as the pancreas. Since this part of the gut is geared to the chemical breakdown of tissue, care must be taken that it does not digest itself. Protection is normally achieved by the secretion of mucus onto the intestinal surface. Mucus also eases the passage of food. In addition to chemically digesting material, the gut is designed to mix, pound and move it along.

Digestion breaks food down into molecules small enough to be absorbed through the gut wall. This uptake of nutrients occurs mostly in the jejunum (the second

part of the small intestine), where the surface area available for absorption is enormously increased by the presence of countless villi – promontories of tissue, like loading wharves, that protrude from the gut wall.

2.36 This section through the human intestinal wall shows it lined by finger-like villi. The 'skin' of the 'fingers' is formed mostly by columnar cells, called enterocytes, with a dark nucleus at their base. These cells have a fast turnover. Formed by rapid cell division in 'crypts' at the bottom of the pits between the villi, enterocytes progress up their sides, only to be shed 4 days later. Tissue full of blood

vessels forms the core of the villi, and runs along their base. The products of digestion are absorbed into this network of vessels, and subsequently transported throughout the body.
LM, haematoxylin and van Gieson stain, ×100

2.37 This electron micrograph provides a view complementary to that of Figure 2.36, showing how a single intestinal villus is constructed. The column-like cells forming its outer surface are interspersed with mucus-secreting goblet cells which, in this specimen, are concentrated in the top half of the villus and are distinguished by the granular secretory droplets they contain. Large blood vessels and smaller capillaries make up the core.
SEM, ×615

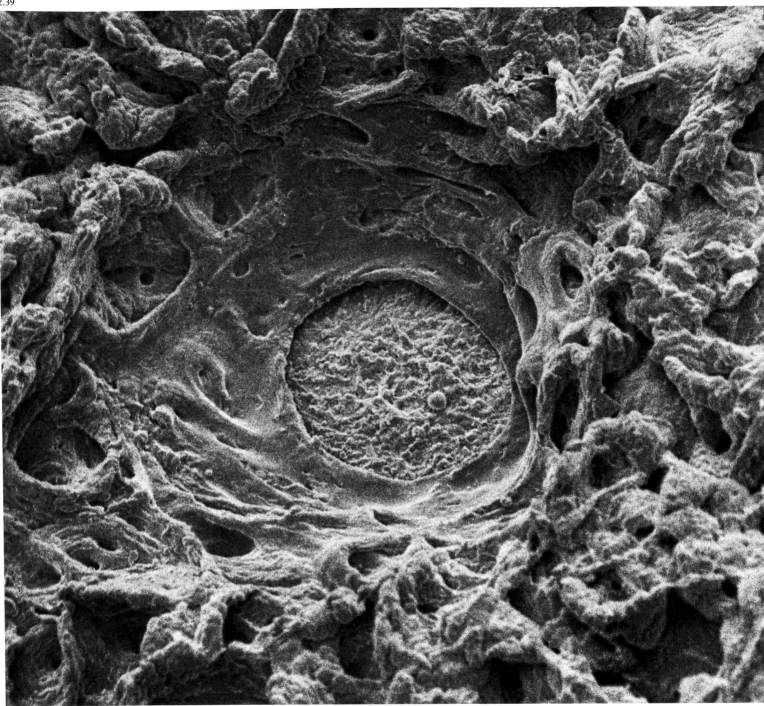

2.38 The orange slug-like creature in the centre of this image is the human intestinal parasite *Giardia lamblia*. *Giardia* are protozoa – single-celled organisms that are the most primitive creatures in the animal kingdom. Here, a *Giardia* uses suckers to attach itself to *micro*villi in the small intestine. Microvilli are tiny tongues of tissue projecting from the outer surface of the enterocytes that line the far larger villi seen in the previous micrographs. Microvilli are of uniform length and so give the impression of a brush-like border. Together, the 30 000 or so on each cell serve to increase the absorptive surface of a villus 30 times. Infestation by *Giardia* (giardiasis) reduces absorption from the intestinal tract and may lead to diarrhoea.
TEM, false colour, ×5000

2.39 With a pH of 2, the juices produced by the stomach are highly acidic. In normal circumstances, the stomach lining is protected by a layer of mucus. But when the mucus barrier is ineffective, the lining itself starts to be digested. The result is an ulcer, as shown in this image of a rat stomach. A characteristic of the body's inner surfaces is that their epithelial cells are generally effective at closing wounds. A sheet of cells moves forward from the borders of the injury, gradually covering the defect and restoring the continuity of the lining. Part of the healing process is seen here. In this instance, the ulcer was artificially induced as part of a research project investigating the mechanisms of wound repair. Such research is important in understanding how human gastric ulcers can be healed more quickly. Although the ulcer shown is a chronic one, and repair is slow, the smooth raised border of epithelial cells is clearly seen advancing over the granular base of the wound. The gastric mucosa surrounding the ulcer is healthy.
SEM, magnification unknown

BLOOD

Blood has two major constituents: cells, and the fluid (plasma) in which they are suspended. The bulk of the cell population consists of red blood corpuscles, or erythrocytes. These cells transport oxygen, and for this purpose are filled with the iron-rich pigment haemoglobin. The remainder are white blood cells, or leucocytes. There are many types of leucocyte, and they are part of the immune system that resists infection. Blood also contains platelets – cell fragments essential to coagulation.

2.40 This smear of human blood consists mostly of erythrocytes. Red cells are concave on both sides, and they are more lightly coloured in the centre because they are thinner there. Mature erythrocytes have no cell nucleus. Two-thirds of circulating white cells are neutrophils, of which two are seen at centre. Neutrophils have a single nucleus, but its division into several lobes often gives a multi-nuclear appearance, as in this case. The lobes are connected by strands of nuclear material.
LM, Giemsa stain, ×675

2.41 The biconcave shape of red blood cells increases the surface area available for gas exchange. Erythrocytes are essentially bags of haemoglobin, composed of nothing but the pigment, a few enzymes, and a highly flexible cell membrane.
SEM, ×6320

2.42 The real work of circulating the blood is done by the 40 000-mile network of capillaries which ensures that no cell in the body is more than a fraction of a millimetre away from a blood supply. Many capillaries, like this one in muscle tissue, are no wider than the red cells which pass along them – as here – in single file. The smallest are only half the diameter of red cells, which must deform themselves to squeeze through.
SEM, ×3740

2.43 The extreme flexibility of red cells is demonstrated by these three

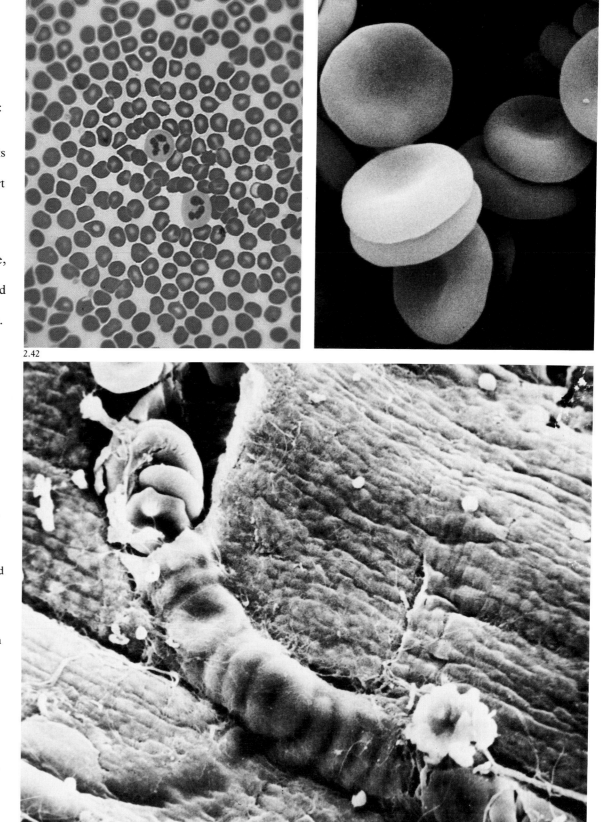

2.40

2.41

2.42

erythrocytes, as they are extruded through a minute opening in a capillary wall (running top left to bottom right). During inflammation, large numbers of white cells and a smaller number of erythrocytes emerge in this way into the tissue surrounding the capillaries. Blood plasma also escapes, contributing to swelling.
TEM, false colour, × 21 750

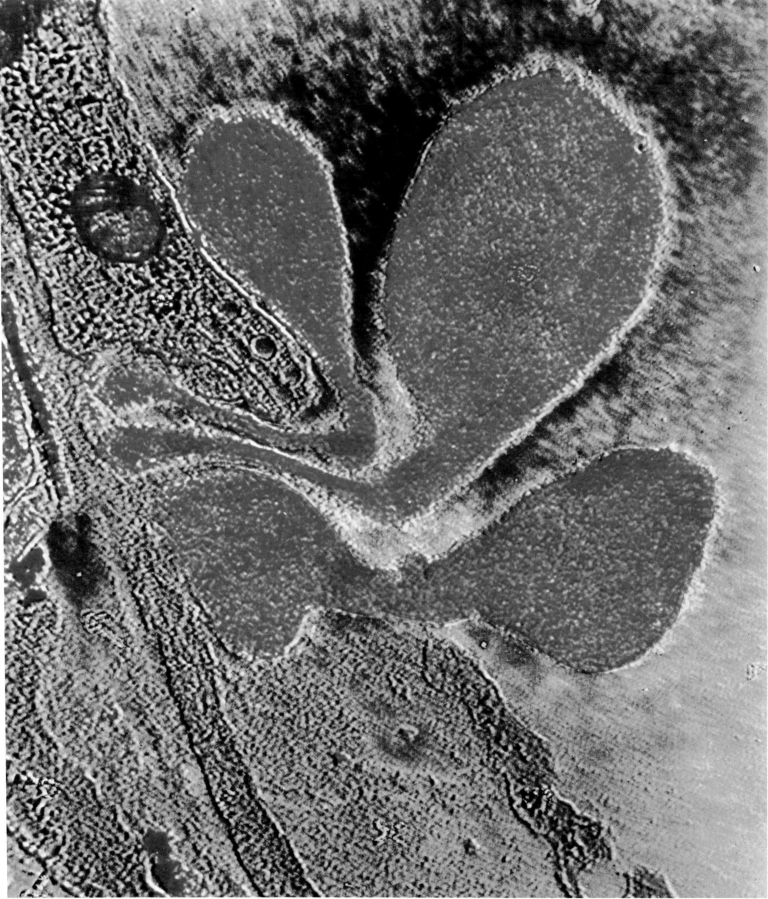

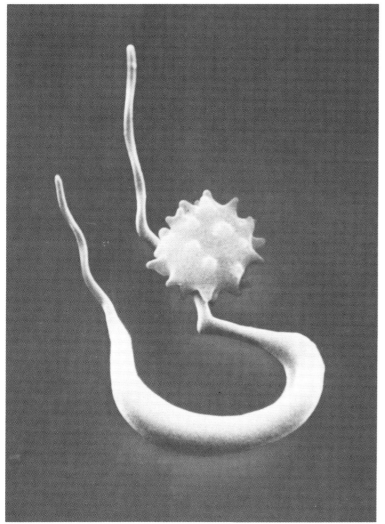

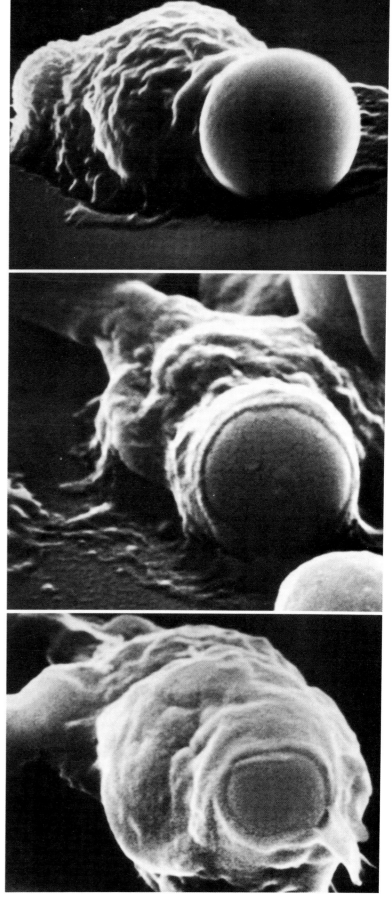

2.44 Sickle cell anaemia is an inherited disease in which the synthesis of haemoglobin is defective. The condition derives its name from a characteristic deformity (shown in the lower cell) of the erythrocytes. Red cells develop this abnormal shape because the haemoglobin molecules form into tiny rods, which assemble into sheaves and stretch the cell membrane. In contrast to the usual flexibility of erythrocytes, sickle cells are rigid. They block fine capillaries, obstructing the blood supply and depriving tissue of oxygen and nutrients. The rounded cell with spines is an echinocyte, another example of the varied shapes red cells can assume.
SEM, magnification unknown

2.45 Erythrocytes are so specialised as transporters of oxygen that they have dispensed with the usual apparatus inside the cell that performs repair. A red cell cannot replenish its enzymes nor mend the wear and tear in its protein coat. As a result, red cells live for only about 120 days. When their useful life is over, they are consumed by white cells. No one is sure what tells the white cell when to move in, although a red cell's inability to maintain its biconcave shape may be involved. Along with neutrophils, white cells called macrophages are the most active scavengers. The process of phagocytosis – literally 'cell-eating' – is dramatically captured in this series of micrographs. A macrophage approaches a red cell, engulfs it in a funnel-shaped 'mouth', and finally swallows it whole. Macrophages are the largest of the body's white cells and are as active against invading micro-organisms and other foreign bodies as they are against old erythrocytes. But phagocytosis is only one element in the body's defence against disease. Other mechanisms involve the immune response, which is described opposite.

SEM, magnification unknown

2.46

2.47

2.48

plasma cells in a mouse spleen are producing two classes of antibody protein: those stained red are secreting immunoglobulin G (IgG), and those stained green immunoglobulin M (IgM). The fact that no cell is stained both red and green shows the specificity of the immune response. Cells specialise in the production of antibodies of one particular sort. The image was obtained by staining tissue from the mouse spleen with two fluorescent dyes. Antibodies to IgM were labelled with the green dye fluorescein, and antibodies to IgG with the red dye rhodamine. The labelled antibodies then bound selectively to cells carrying the corresponding immunoglobulin on their surface.
LM, fluorescence, ×950

2.47 When an antibody combines with its antigen, it can trigger the destruction of the cell to which the antigen belongs in a number of ways. One is by activation of 'complement' – a system of nine proteins, designated C1–C9, which circulate in blood plasma. The tree-shaped protein molecule in this extremely high-magnification transmission electron micrograph is termed C1q. An antibody attached to a target cell will bind to one of C1q's six 'heads'. This triggers a cascade of the other complement proteins, from C1 to C9, which causes the punching of holes in the membrane of the target cell, leading to its destruction.
TEM, negative stain, ×1 400 000

2.48 As well as being activated by invading organisms, the immune system responds to any of the body's own cells that are abnormal, such as tumour cells. This scanning electron micrograph shows four small T-lymphocytes (so called because they mature in the thymus gland) attacking a large cancer cell. Some T-cells produce substances that attract patrolling macrophages and stimulate their phagocytic activity. Others – like these – attack the target cell directly and are known as killer T-lymphocytes. Unfortunately, as we know from the prevalence of cancers, their efforts are not always successful.
SEM, ×1125

IMMUNE SYSTEM

Bacteria and other organisms that penetrate our outer defences contain large proteins (antigens) which the body recognises as foreign. One way we respond is by producing antibodies – specific proteins which circulate in the blood until they can combine with their target antigen. Manufacturing antibodies is the job of B-lymphocytes. Lymphocytes are white cells which mature in tissues of the lymphatic system such as the thymus and spleen. Other kinds of lymphocyte – the T-cells – act more directly to destroy invading organisms.

2.46 B-lymphocytes that encounter antigens divide to form antibody-secreting 'plasma' cells. Such cells are capable of secreting 2000 identical antibody molecules per second for the few days of their mature life. In this immunofluorescence micrograph,

SKIN & HAIR

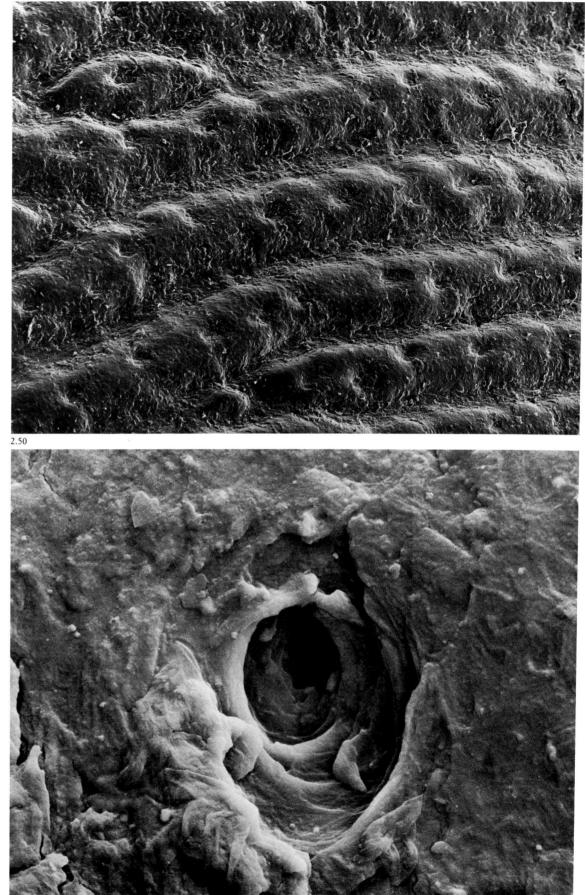

Although only a few millimetres thick, our 25 square metres of skin form the body's largest organ, and an important interface with the world. Because it is only sparsely covered with hair, human skin is vulnerable. But the 'horny', cornified, outer layer – the epidermis – is some protection; and when intact the skin is an effective barrier against viruses and bacteria. Below the epidermis is a dense layer of tissue – the dermis – rich with blood vessels and nerve endings. The skin is a versatile sensory organ, responsive to touch, pressure, pain, heat and cold; and vital in temperature regulation.

2.49 Pronounced valleys and ridges are found in skin from the palm of the hand (shown here), the soles of the feet and the fingertips. The precise pattern is unique to each of us. Sweat pores appear as small craters along the ridges. Cells nearest the skin surface – the cornified layer – have been flattened and hardened by deposition within them of the protein keratin. This tough, dead barrier of cell remnants is continuously being shed, and its flakiness is evident in the micrograph. Dead cells are constantly replaced by newer, maturing cells, which take a month to migrate from the base of the epidermis.
SEM, ×30

2.50 Like the opening to a deep cave, a sweat pore spirals down through the outer layer of the skin. It ends in a coiled sweat gland in the dermis or subcutaneous layers. This pore is on the palm of the hand – one of the areas of the body richest in sweat glands. One quarter of the body's heat loss is achieved through sweating.
SEM, ×560

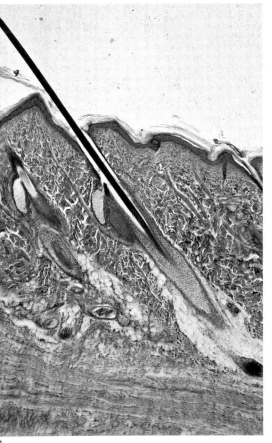

2.51

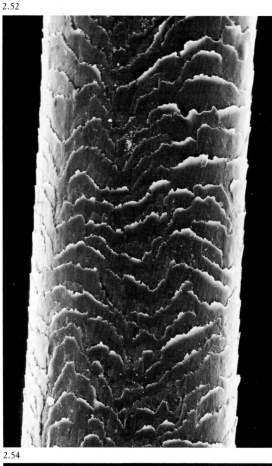

2.52

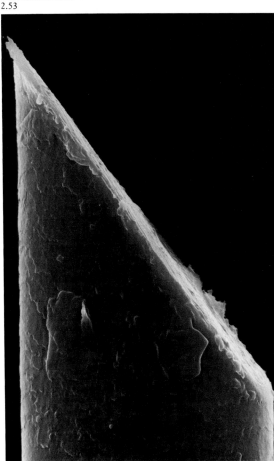

2.53

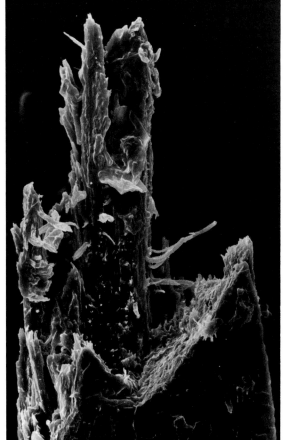

2.54

2.51 The different layers of the skin, and a single hair, are shown in section in this light micrograph. The horny outer layer of the epidermis is seen as the thin, darker band, supported by the much deeper but less dense dermis, which merges into the subcutaneous layer at the bottom of the picture. Note how the epidermis grows down to form the lining of the hair shaft, or follicle, below the skin surface. The hair is rooted in an enlarged portion – the hair bulb, a region of actively dividing cells from which the hair grows. Hair may curl when the follicles are curled, or when the hair bulb lies at an angle to the shaft. Associated with each follicle is a bundle of smooth muscle, responsible for pulling the hair erect in conditions of cold or fear, and one or more sebaceous glands that secrete the oily, waterproofing agent sebum onto the hair and the skin surface. A sebaceous gland can be seen half way down the left side of the hair follicle.
LM, magnification unknown

2.52 Hair consists largely of keratin. This scanning electron micrograph of the surface of a normal human hair clearly shows the overlapping keratin plates, or scales, which are thought to reduce hair matting.
SEM, ×480

2.53–2.54 This is the kind of comparison that infuriates the manufacturers of electric razors. On the left is a man's beard hair, cleanly sliced by the blade of a 'wet' razor. On the right is a beard hair from the same man, torn and mangled by the action of an electric razor. Since each hair is approximately one fifth of a millimetre in diameter, the difference visible to the naked eye is little or none, and the difference felt by someone kissing is likely to be more imagined than real. Nonetheless, one can see why the manufacturers of electric razors would have preferred it if the scanning electron microscope had never been invented.
SEMs, both ×275

CHAPTER 3
ANIMAL LIFE

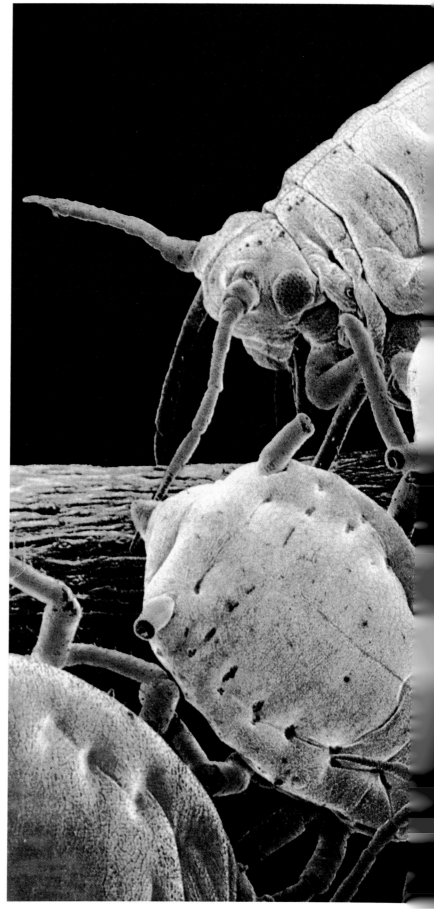

MICROSCOPES serve to counteract our biased view of the animal kingdom. The mammals, birds, reptiles, amphibia and fish which loom so large in our consciousness disappear altogether. They are replaced by the protozoa, worms and arthropods covered in this chapter. In terms of both the number of species and the number of individuals, it is these creatures which truly dominate the animal life of Earth.

The most important characteristic defining a creature as an animal is the way in which it obtains energy for growth. Most plants and many micro-organisms need only simple minerals, water, and usually air and light. Animals require preformed food from an organic source. This requirement dictates many aspects of their lifestyle, such as the development of sensory systems and locomotion to enable them to catch prey and avoid becoming prey themselves. Feeding and digestive mechanisms are adapted to the food source. Insects have a great variety of chewing, cutting, or piercing mouthparts for eating wood, leaves, or blood. Tapeworms have no mouth and no digestive system; they inhabit the intestines of other animals, where they absorb the already digested food of their hosts. The degree to which the body of an animal has specialised in directions like these provides a basis for classification.

The most 'primitive' animals have relatively poor senses and a simple body structure consisting of just one cell, or a colony of single cells joined together. Protozoa nonetheless perform all the functions of an animal within their single cell, which can be extremely complex. 'Higher' animals have a multicellular body with a nervous system and specialised organs and tissues; the cells which comprise them usually perform one or a handful of specific tasks, and may be comparatively simple.

Important steps in the evolution of 'higher' animals were the development of a special internal body cavity or 'coelom' (in the roundworms); the development of separate body segments (in the annelid worms); and the acquisition of a protective cuticle or exoskeleton (in the arthropods). Taken together, they made possible the great range of structures and lifestyles of the arthropods in general and the insects – the most successful of all animals – in particular.

Amateur and professional light microscopists have long delighted in observing small aquatic animals, from protozoa to the larvae of insects and crustacea. Many move with great speed and elegance, seeking out food, escaping danger, or reproducing as one watches. The scanning electron microscope cannot show us a living world, but its detailed images of small animals reveal how complex they can be.

3.1 The aphids (Aphidoidea) are a large, successful group of insects. Commonly known as plant lice, they are specialised plant feeders that exploit two different hosts during a season. This group of unidentified wingless aphids is feeding on a plant stem. The proboscis of the aphid at far right has penetrated the stem and is drinking the sugary juices within. Large populations of aphids develop rapidly and plants are quickly depleted of food, causing them to wilt and die. At the posterior end of each aphid are two short tubes, called cornicles, which secrete pheromones and wax. At the front end are the segmented antennae, positioned in front of the lateral compound eyes.
SEM, false colour, ×175

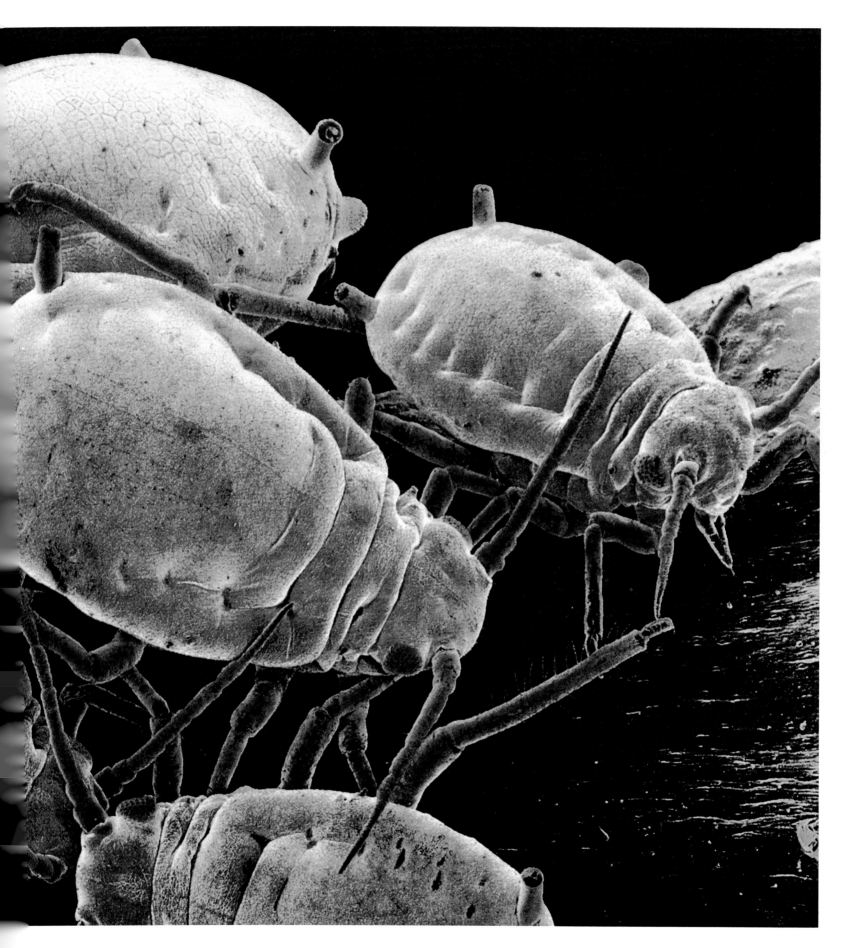

PROTOZOA

The protozoa, or 'first animals', are by definition located at the start of the animal kingdom. They are its simplest representatives, and the vast majority consist of a single cell with all the equipment necessary for an animal existence. A few species classed as protozoa are exceptional, in that they obtain energy from sunlight in the same manner as plants. Some 80 000 species have been described and these are divided into four classes: the flagellates, amoebas, ciliates and sporozoa.

Protozoa have not only the basic cellular organelles such as a nucleus and mitochondria, but also an array of additional structures that enable them to feed, excrete waste, protect themselves and move around the environment. Some species form an external protective layer called a pellicle, and some a complex shell or test.

Most protozoa feed like animals by breaking down organic compounds and absorbing their constituents. Many have a 'cell mouth', which among the ciliates is an elaborate, fixed structure, while in some amoebas it is temporary and disappears once feeding ceases. As the food reaches the animal's cytoplasm, it is enclosed in a food vacuole, where it is digested by the action of enzymes.

Several mechanisms are used by protozoa for moving about their environment. The flagellates use flagellae – whip-like filaments that cause an undulating, sometimes gyratory, motion through the water. Ciliates rely on cilia, or fine hairs, which beat against the water, creating a current. Amoebas advance using pseudopodia, or 'false feet'. These are extensions of their cytoplasm, which they channel in one or

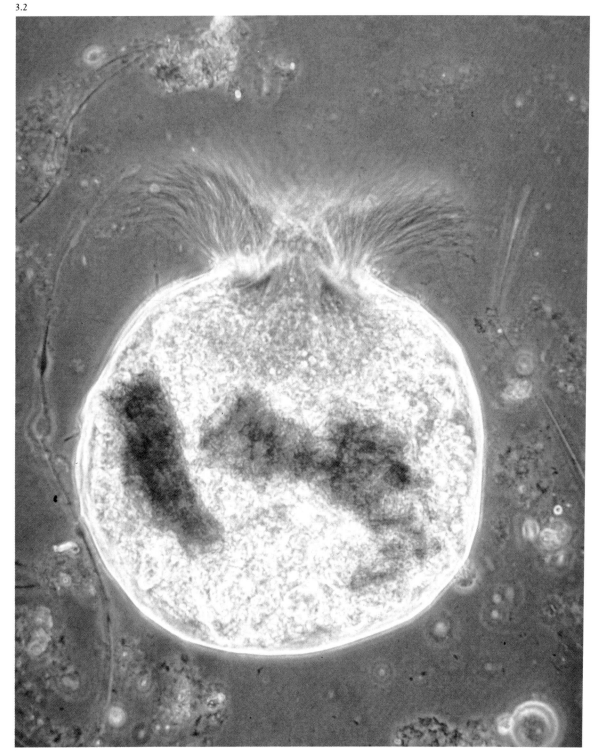

several directions simultaneously.

Protozoa are found in freshwater, seawater and in abundance in the soil. The majority form a valuble link at the base of the food chain. Some species, however, are parasitic and a few of these cause serious disease in humans. The malarial parasite *Plasmodium* is a protozoon, and so is *Entamoeba hystolytica*, the cause of amoebic dysentery.

3.2 The flagellate *Barbulanympha ufalula* forms a symbiotic relationship with a rare wood-eating cockroach, *Crytocercus puntulatum*. It lives in the hindgut of the cockroach, where it provides the essential digestive enzymes lacking in its host. The cockroach feeds exclusively on wood, but is unable to convert cellulose into carbohydrate. In return for doing this, *B. ufalula* shares in the digested products. The cytoplasm contains two brown fragments of half-digested food, above which is the multiflagellate crown, concealing the mouth area. LM, ×520

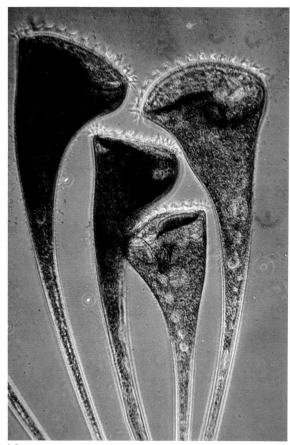

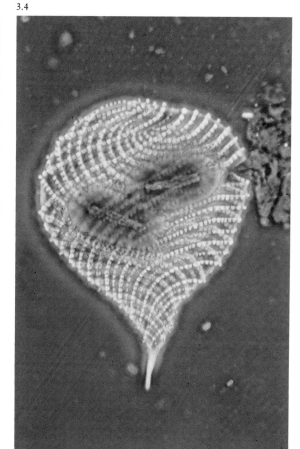

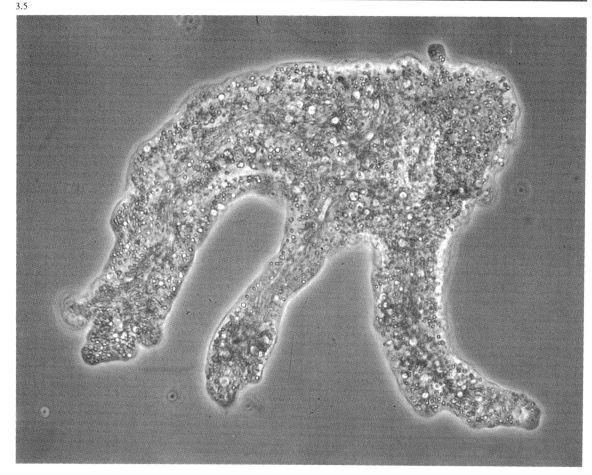

3.3 The trumpet-shaped freshwater ciliate *Stentor coeruleus* swims freely or attaches itself to a substrate – a filament of algae, for instance – by using its tapering 'foot' as an anchor. *Stentor* is highly contractile, shrinking down close to the substrate when disturbed or irritated. This light micrograph shows a group of *Stentor* extended in a feeding position. The broad lip of the bell-shaped mouth and the surface of the cavity are covered with rows of compound cilia, which beat rapidly, drawing in and filtering a variety of microscopic food particles such as bacteria, diatoms and rotifers. LM, ×400

3.4 The pellicle, or outer layer, provides protozoa with a degree of rigidity. Its structure varies considerably between species. The specimen in this micrograph, *Euglena fusca*, is a flagellate. It was punctured and its cell contents squeezed out (they can be seen at top right) to give a clear view of the pellicle. Two brown paramylon bodies, which store surplus carbohydrates, remain within the cell membrane. The pellicle is seen to consist of a series of overlapping spiral striations made from a protein and strung with beads of mucilage. The beads are impregnated with iron, giving the animal a pink tinge. LM, ×960

3.5 Amoebas are nature's doodlers, constantly changing their body shape in order to move and feed. Some species only produce one pseudopodium, but this specimen has three; they are seen extending towards the bottom of the image. Just how pseudopodia are formed is still a mystery. The cytoplasm of amoebas is divided into two zones: an inner, fluid region called the endoplasm, which consists of most of the visible cytoplasm, and an outer, gel-like layer lying under the cell's outer membrane. The interaction of these zones is thought to somehow control the protoplasmic streaming that produces the pseudopodia. LM, magnification unknown

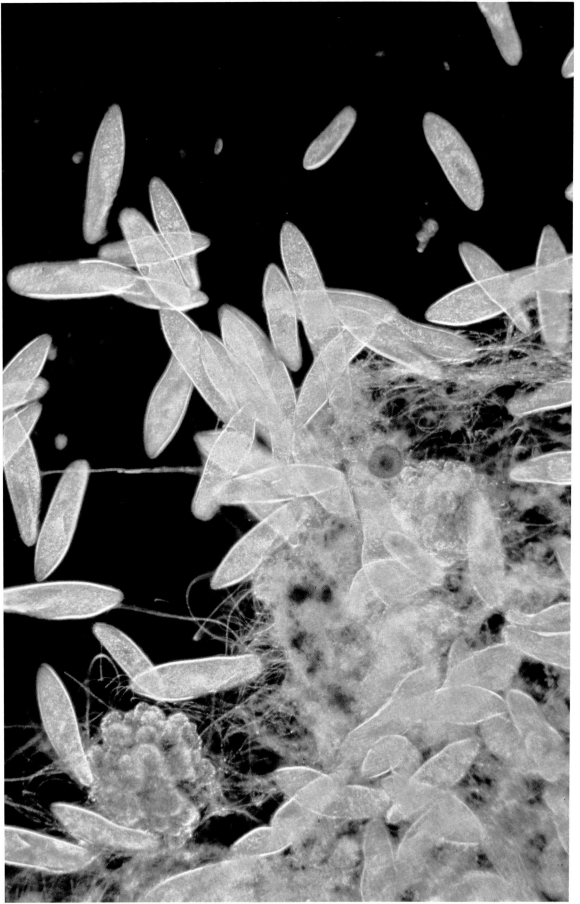

3.6 A drop of pondwater observed through the light microscope becomes a crowded microworld. Here a group of the slipper-shaped ciliate *Paramecium sp.* is browsing among a clump of filamentous bacteria (*Beggiatoa sp.*). *Paramecium* are abundant in water and soil. They have a permanent mouth, and a varied diet which includes algae, diatoms and decaying organic matter.
LM, dark field illumination, ×160

3.7–3.10 *Paramecium* is a favourite food for many organisms, but none so bizarre as fellow ciliate *Didinium nasutum*. This sequence of scanning electron micrographs shows a *Didinium* attacking and consuming a *Paramecium* almost twice its size, in just one mouthful. *Didinium* is barrel-shaped, with two girdles of locomotory cilia and a prominent snout. The snout is used as a probing and seizing organ during feeding. It is reinforced by numerous rod-like structures, which are used to hold fast the prey. In Figure 3.7, the *Paramecium* senses the approaching threat and ejects its arsenal of trichocysts, a tangle of fibrous rods which scare some predators but have little effect on *Didinium*. In Figure 3.8, the *Didinium* grips the *Paramecium* near the latter's mouth; the *Didinium's* snout and gullet expand enormously in preparation for the meal. In Figure 3.9, the *Didinium* manoeuvres the *Paramecium* into an easier position for ingestion. In Figure 3.10, the *Didinium's* seizing organ or snout moves back through the interior of the cell, drawing the *Paramecium* into the body cavity along with it. The *Paramecium* completely fills the interior of the *Didinium*, which now resembles a giant food vacuole. Once digestion is completed, the snout returns to the front of the cell in position to feed again. Under laboratory conditions a *Didinium* can consume a *Paramecium* every three hours.
SEMs, all ×700

3.7

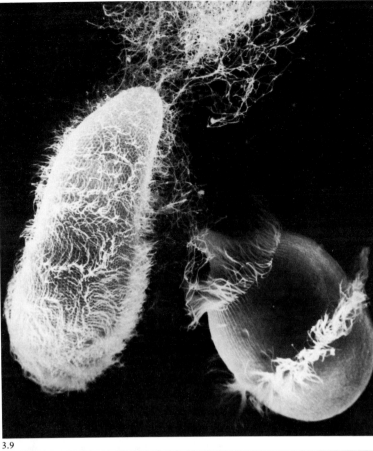

3.8

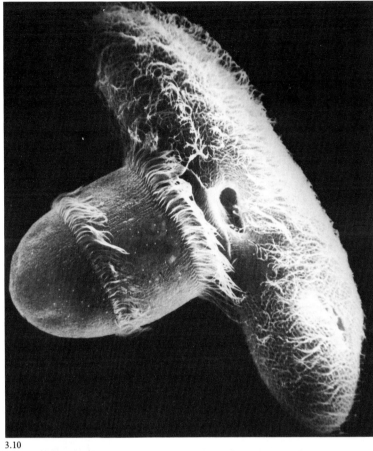

3.9

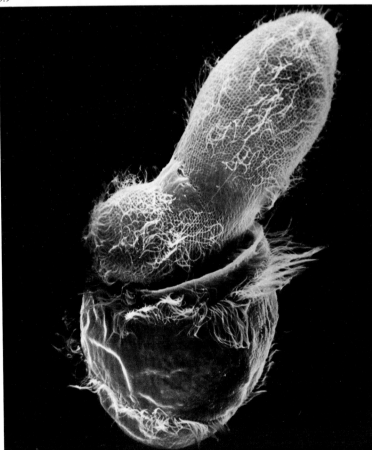

3.10

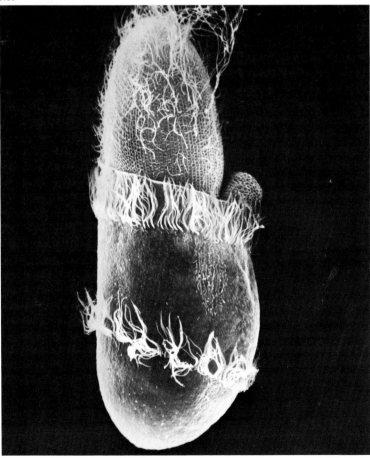

PARASITIC WORMS

Many organisms lead a parasitic lifestyle, but few have developed it to such a peak as the parasitic worms. The predominant groups are the trematodes or flukes, the cestodes or tapeworms, and the nematodes or roundworms. Between them, they live on or in almost every type of creature: insects and molluscs, reptiles, birds and mammals. There are about 70 species which live in humans: the 10 most common species infect *half* the world's population, though only a minority of those infected develop serious disease symptoms.

A marked characteristic of parasitic worms is the complexity of their life-cycle, which usually involves two or more host species at different stages in a worm's development from larva to adult. The fluke *Alaria mustelae*, for example, inhabits in turn a species of aquatic snail, a frog and a rodent; it then matures and reproduces in its primary or 'definitive' host, the North American mink. Worms such as *A. mustelae* can only mature in one specific host. Others are less particular; the blood fluke *Schistosoma japonicum* can mature in cats, dogs, rodents, pigs, sheep, goats, cattle and humans. Such versatility makes it all the more difficult to eradicate.

The complex life of parasitic worms is remarkable in view of their 'simplicity' in evolutionary terms. They are altogether more primitive than the most familiar worm, the common earthworm. The latter is a segmented animal and, apart from its worm-like appearance, is more closely related to crustacea and insects than to the parasitic worms. Flukes and tapeworms belong to the phylum of flatworms or Platyhelminthes, and they are the simplest multicellular animals with bilateral symmetry. They have no body cavity or coelom, no vascular system, no respiratory system and no anus. A cluster of nerve cells at the front end of their body serves as a primitive 'brain'. The nematodes belong to the phylum of roundworms or Aschelminthes, and are a little more developed – for instance, they have a coelom and an anus.

3.11 The head or 'scolex' of the small tapeworm *Acanthrocirrus retrirostris*, seen here in its larval form, includes a piston-like apparatus which can be withdrawn into the scolex or thrust out and buried in the tissue of its host. The hooks decorating its tip operate like the ribs of an umbrella when opened and closed. Beneath the hooks, staring out like alien eyes, are two of the four suckers grouped around the scolex; they are a further means of attachment to the host. The tapeworm larvae parasitise the barnacle *Balanus balanoides*; they reach maturity in the intestines of wading birds such as the turnstones, which feed on the barnacles.
SEM, ×475

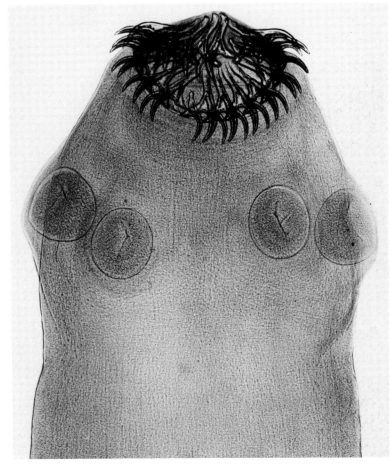

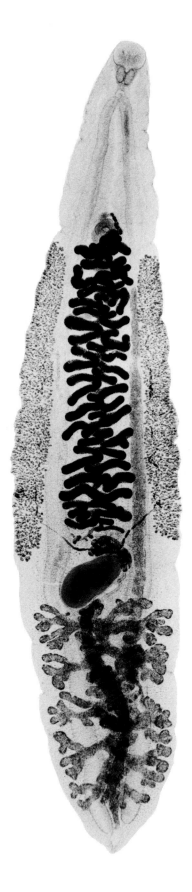

3.12 An estimated 4 million people are hosts of the pork tapeworm *Taenia solium*, whose four suckers and hooked scolex are seen in this light micrograph. Adult tapeworms always live in their hosts' intestines, surrounded by digested food which they absorb directly through their entire surface area – relieving them of the need for a digestive system of their own. Tapeworms grow a series of hermaphroditic reproductive organs, one after the other, and it is these well-defined 'proglottids' which make adult tapeworms look as if they are segmented. Each developed proglottid may contain thousands of eggs. Ripe proglottids are detached from the end of the worm's 'ribbon', and released into the environment in human faeces. It is at this stage that they are liable to be ingested by domestic pigs, in which the eggs develop into larvae. Pigs are the worm's main secondary host, but the larvae can live in many other animals. Humans can be secondary as well as primary hosts, and in such cases the larvae tend to migrate to the nervous system or eyes, where they can cause severe damage.
LM, ×50

3.13 This micrograph shows the internal structure of the liver fluke *Clonorchis sinensis*, which infects some 28 million people in China, Korea and Japan. It can cause diarrhoea, abdominal pain and cirrhosis of the liver. Flukes are usually shorter than tapeworms, and oval in shape. They also have a digestive system, seen here as the faint pair of intestinal tubes running from the oral sucker at the animal's front end (top) down the entire length of its body. The branched structures at the bottom of the picture are the hermaphroditic worm's testes. They deliver sperm into the sperm store, the dark grey oval at lower centre. The small, dark, irregular body nestling around the sperm store's narrow end is the ovary. The black 'fingers' at upper centre are the uterus, on either side of which are the granular vitelline glands. These contribute to the development of the worm's eggs. Like most flukes, *C. sinensis* has two secondary hosts – first, freshwater snails; then, freshwater fish of the minnow and carp family. Humans are infected by eating raw fish.
LM, ×15

3.14 Blood flukes are so called because the adults live in the veins of their hosts. Unlike other flukes, which are all hermaphrodites, blood flukes have separate sexes and the adults live as couples, the female lying within an abdominal groove in the male. In this light micrograph, the male is thicker and blue-coloured, the female white and thread-like. The brown ribbon within the female is half-digested blood from her last meal. Parts of other worms are visible at left and bottom right. There are three medically important species of blood fluke: *Schistosoma mansoni* (shown here) and its relatives *S. haematobium* and *S. japonicum*. They infect some 200 million people in Asia, Africa and Latin America. A person may be host to several dozen pairs of worms, each of which produces thousands of eggs. The disease of schistosomiasis, also known as bilharzia, results primarily from the eggs accumulating in tissue and causing internal bleeding, anaemia, dysentery and other symptoms. Schistosoma eggs released via the faeces or urine find their way into the water, where they hatch into larvae. The larvae infect freshwater snails which are the fluke's secondary hosts. Inside a snail, the first-stage larvae reproduce asexually to generate large numbers of second-stage larvae which swarm out of the snail and into the water. It is at this point, in irrigation ditches, rice paddies, streams and rivers, that people become infected. The larvae bore into a person's skin until they find a vein; next they travel through the venous system, via the right side of the heart and lungs, until they reach the liver, where they mature and seek a mate with whom to live in their host's veins.
LM, dark field illumination, ×30

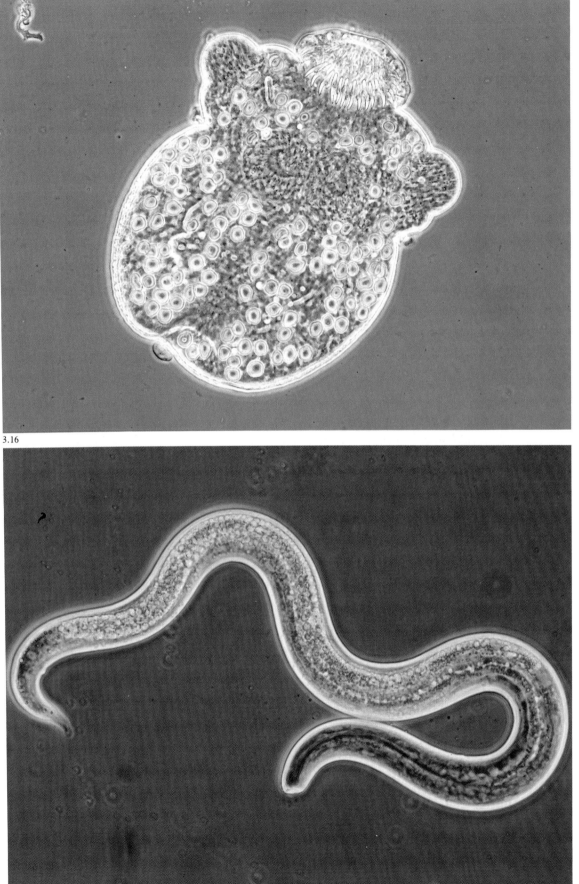

3.16

3.15 This light micrograph shows the larval form of the dog tapeworm *Echinococcus granulosus*. Although the adults parasitise only dogs, the larvae are at home in ruminants, rodents, pigs and horses, and occasionally humans. They cause the rare but dangerous disease of hydatidosis, named after the 'hydatid' cysts, containing numerous larvae, which form in the liver, lungs, brain, or other organs. The nature of the disease depends on which organ is affected. In the brain, the cysts can cause blindness and epilepsy. When the larvae emerge from the cyst, they migrate to the intestines, where they use their crown of hooks, visible in this picture, to secure them to the gut wall. The numerous circular bodies inside this larva are 'calcareous corpuscles'. They consist of an organic base and inorganic materials such as calcium, magnesium, phosphorus and carbon dioxide. Their role is uncertain, but they may act as a source of energy.
LM, ×355

3.16 Dogs are also the definitive host of the nematode *Toxocara canis*, the first larval stage of which is seen in this micrograph. Nematodes have four larval stages before they mature into adults. Humans can become infected by the eggs and larvae of *T. canis*, and although these do not take up permanent residence inside us, they cause tissue damage as they migrate around the body. Some 80 000 species of nematodes have been described, but at least 10 times as many are thought to exist. They are found in vast quantities as free-living marine, freshwater and soil organisms, and also as animal and plant parasites. The smallest is only 200 micrometres long; the largest grows to several metres and parasitises the placenta of sperm whales.
LM, ×570

ROTIFERS

Although they are completely different in appearance, rotifers also belong to the phylum of Aschelminthes, the roundworms, which includes the nematode on the previous page. Ranging from 0.04 to 2 millimetres in length, and present in almost any sample of freshwater, rotifers have delighted generations of light microscopists with their beautiful and varied forms, their constant activity and the transparency which makes it possible to distinguish their internal organs. The early microscopists called them 'wheel animalcules', or rotifers, because the synchronised beating of the cilia that encircle their funnel-shaped mouths resembles the spinning of tiny wheels.

There are some 2000 species of rotifers. Some are marine, but most live in freshwater, on waterlogged mosses, or in the soil. If their environment dries up, many rotifers form cysts in which they can survive for years.

The vast majority of rotifers are females. In many species, males are unknown and reproduction is thought to occur solely by 'virgin birth', or parthenogenesis. In others, males occasionally occur, but they are unable to feed and live for only a few hours or days; most of their organs are degenerate, except for those which enable them to locate and mate with females.

3.17 This specimen of the common pond-dwelling rotifer *Philodina sp.* is seen feeding among a tangle of filamentous and other algae. Its funnel-shaped head and mouth are tipped with the beating cilia which create currents that waft organic detritus, algae, bacteria and protozoa into its jaws.
LM, Rheinberg illumination, ×160

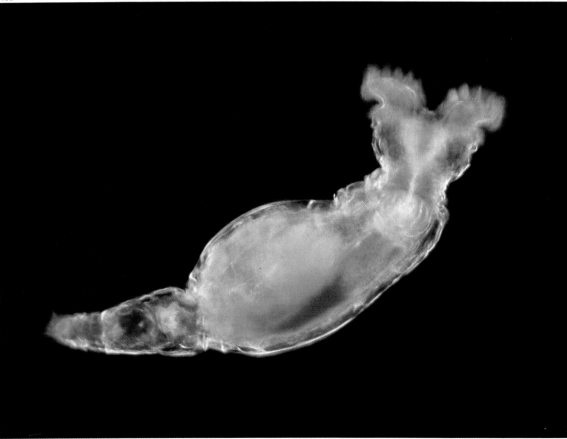

3.18 *Philodina gregaria* is an Antarctic species which survives the polar winter in cyst form and returns to activity when pools and lakes of meltwater form in the brief austral summer. It can then reproduce in such vast numbers as to colour the floor of a lake red. The picture shows clearly its paired ciliary organs and its Y-shaped mouth and gullet.
LM, dark field illumination, ×220

3.19 The three rotifers in this micrograph are feeding on a filamentous green alga. The larger rotifer is a member of the genus *Mytilina*. Its mouth is at left, together with its red eye-spot. The thorn-like projection at the other end of its body is one of a pair of long toes; the other toe is out of focus behind the first. Like a number of other rotifers, *Mytilina* is protected by a kind of transparent 'shell', known as a lorica, which surrounds the main part of its body. The green area within the rotifer is its stomach, its contents coloured by ingested algae; the white area is its vitellarium or egg-producing organ. The oval immediately below *Mytilina*'s head is an unidentified rotifer egg. The two smaller rotifers in the top half of the picture, also equipped with distinctive toes, belong to the genus *Lecane*.
LM, dark field illumination, ×160

3.20 *Floscularia ringens* is a relatively large rotifer and is one of the sessile species which attaches itself to a permanent base – a piece of waterweed in this case. Its ciliated lappets waft towards it particles of mud which it manufactures into tiny mud balls. It adds them in a defined pattern to form the tube with which it surrounds itself and into which it withdraws when danger threatens. *F. ringens'* tube can itself act as a base for smaller sessile rotifers with their own tubes (into which the one at the left has withdrawn), and for rotifer eggs like the three at upper right.
LM, dark field illumination, ×50

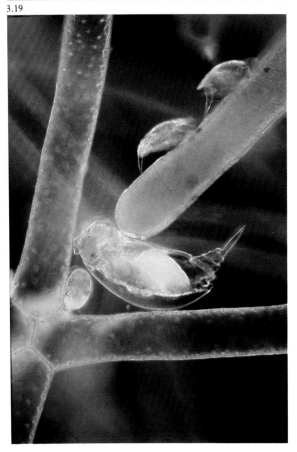

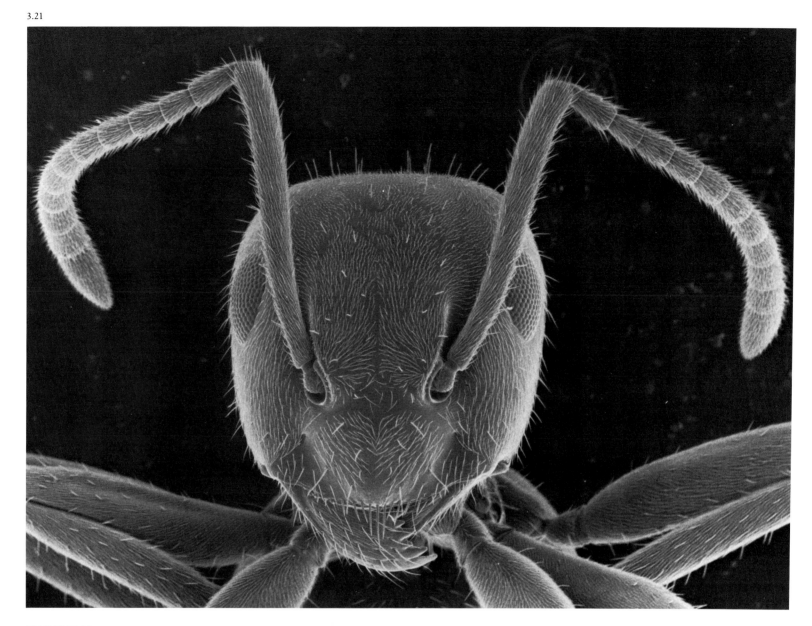

INSECTS

Insects are by far the largest group in the animal kingdom. Over one million species are known and many of these have enormous populations of individuals. This numerical superiority is matched by a bewildering diversity of structure and habitat, reflecting the insects' extraordinary adaptability.

A contributing factor to the success of the insects is their hard outer skeleton, made of chitin.

This versatile suit of armour protects insects from drying out, an adaptation which enabled them to leave the moist environment required by their worm-like ancestors and spread onto dry land. With the development of wings, insects colonised the air.

The process of metamorphosis, experienced by most insects to a varying degree, is not only a source of wonderment to anyone who witnesses it, but an important adaptive device for the animal. The separation of the larval stage, both in structure and habitat, from the adult stage ensures that the two forms do not compete for food supplies. Most insects experience three or four stages in their life-cycle: egg, larva, pupa and adult.

Adult insects are divided into three parts: head, thorax and abdomen. The head carries most of the sensory equipment – compound eyes and antennae studded with tactile and olfactory organs. The mouthparts are varied according to the diet of the species. Beetles, ants and termites feed on solid matter such as wood and seeds; they are equipped with powerful mandibles for grasping, chewing or cutting. Butterflies, aphids and flies have a sucking or piercing proboscis to suit their liquid diet of nectar, plant sap, or blood.

The thorax of the adult insect carries the six legs and two pairs of wings. Most adult insects fly, but by no means all. The springtails have never developed wings and the sucking lice have dispensed with theirs; among the ants, the queens and appointed males fly once during the nuptial flight, but the workers and soldiers never fly at all.

Humans live uneasily with insects. Some species are simply a nuisance, others are common pests, particularly in agriculture. The microscope has confirmed that some species are carriers of disease-causing micro-organisms. While feeding on human blood, the *Anopheles* mosquito transmits the malarial parasite and the tsetse fly the agent of sleeping sickness.

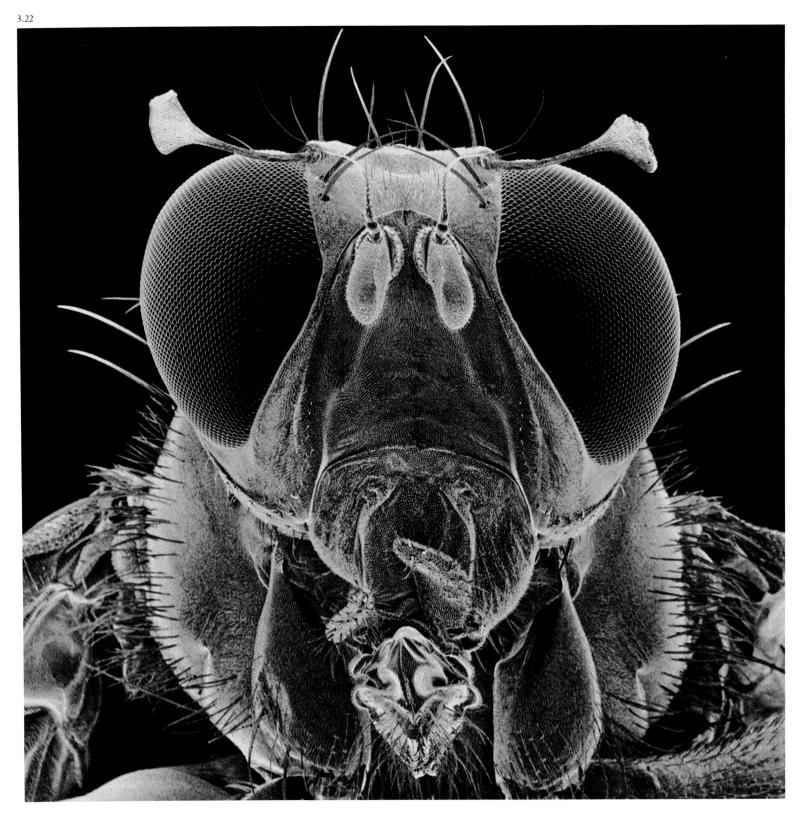

3.21 The head of an insect is a tough capsule welded together from six segments. The mouthparts of the black garden ant *Lasius niger* are of the chewing variety. The mandibles, at the bottom of the head of this female worker, terminate in large teeth, which grasp the food and pull it apart. Ants use their antennae constantly to gather information. The antennae are segmented and attached to the head by a ball-and-socket joint.
SEM, ×60

3.22 Large compound eyes dominate the head of the Mediterranean fruit fly *Ceratitis capitata*. The 'flags' above the eyes are male features. The reduced antennae, situated between the eyes, consist of two stubby protuberances decorated with bristle-like extensions called aristas. A pair of palps, or tactile feelers, are positioned above the trumpet-shaped opening of the proboscis at the bottom of the picture. The proboscis is a sucking tube which ends in two fleshy lobes called labellae. These mop up fermenting juices through a system of fine tubes. The fruit fly is a serious pest to citrus growers.
SEM, false colour, ×115

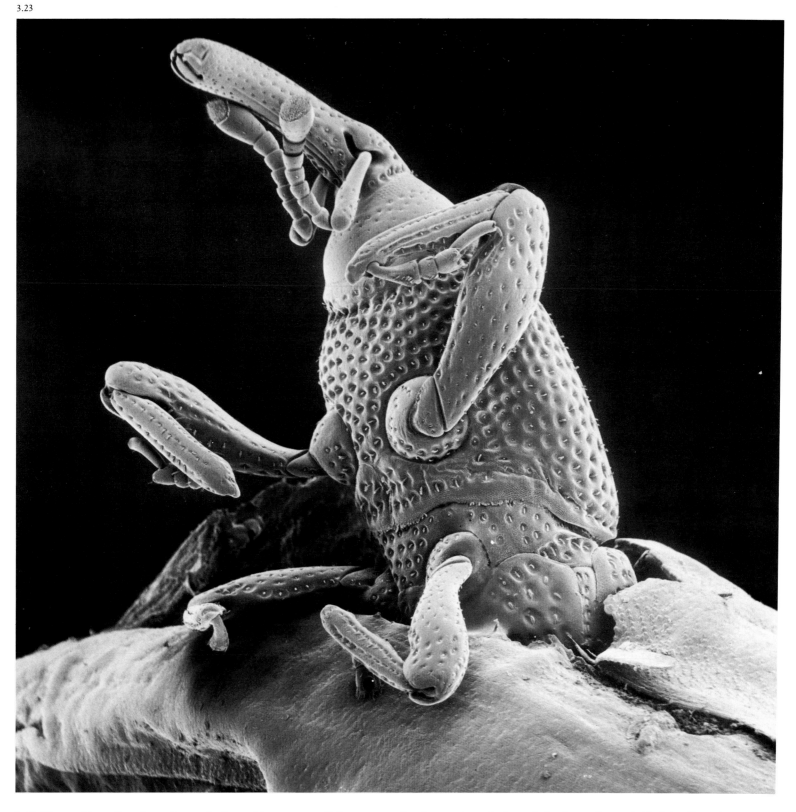

3.23 The graphic quality of scanning electron micrographs confirms the view of many that insects are miniature monsters, unpleasant and to be avoided. As this micrograph shows, it can also reveal a humorous and endearing character. The grain weevil *Sitophilus granarius* is a beetle. Its head is elongatd into a snout called a rostrum, which carries at its tip the blade-like mandibles. The rostrum is used to bore through the fibrous coat of a wheat grain and the mandibles are used for chewing and crushing the kernel. Female weevils insert their eggs in cavities dug out of the kernel, thereby ensuring a food supply for the larvae. Stored grain is particularly vulnerable to infestations of *S. granarius*, although its importance as a pest has declined with the use of fumigants and better storage methods. The compound eye is barely distinguishable as a clump of small dots set behind the elbowed antennae. The first thoracic segment of beetles is enlarged to form the 'pronotum'. The design of the pronotum varies; the weevil's fits like a bulletproof vest at the base of the head and encloses the first pair of legs.
SEM, ×50

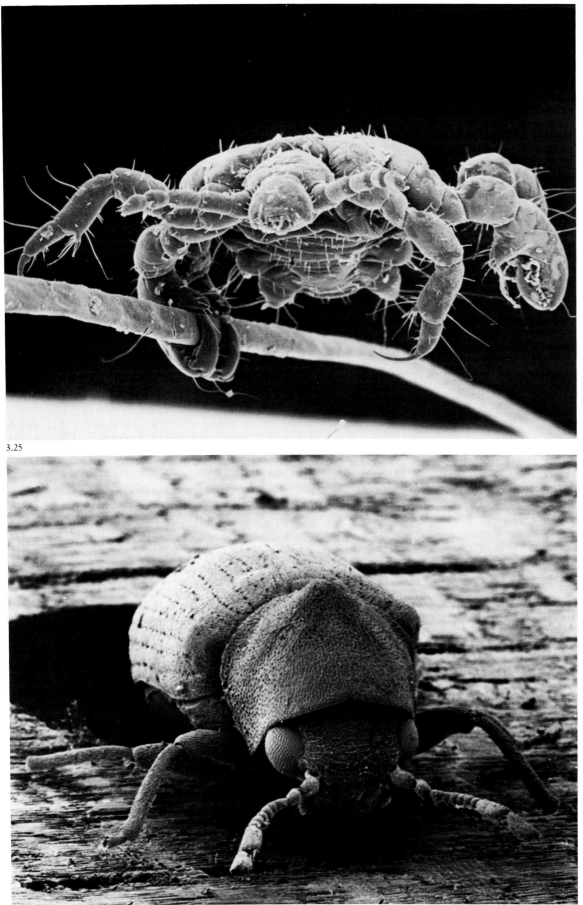

3.24 With its claws firmly gripping a pubic hair, the pubic or crab louse *Phthirus pubis* is a fearsome beast. It belongs to the group of blood-sucking lice, all of which are parasitic on mammals. *P. pubis* is an external parasite on humans, choosing the warm regions in the pubic hairs as its feeding grounds. An infestation of lice causes pediculosis, the symptoms of which are severe itching and a rash. Lice spread rapidly by direct contact and through bedding and clothing. Each of the louse's six legs terminates in a massive claw, which folds inward to meet a thumb-like projection on the opposite side. The louse climbs and swings through its environment, locking into position when disturbed. The piercing proboscis forming the mouthparts is small and shows indistinctly in this micrograph. It consists of three slender stylets, one of which ends in four toothed processes used for sawing into the skin. A louse feeds five times a day, taking about 35–45 minutes to drink its fill. The stout antennae are visible on each side of the head.
SEM, ×80

3.25 Another pest, dreaded this time by the property owner, is the woodworm beetle *Anobium punctatum*. Its larvae cause the most damage; they feed on the dead wood of trees, or rafters, beams and household furniture. The adult supplements its wood diet with pollen and nectar. On hatching, the larvae tunnel into wood, chewing a path of destruction through its interior. The adult insect, seen emerging from a wood hole, uses toothed mandibles below its antennae to shred the wood. Woodworm damage can be extensive without being immediately noticeable. The few holes on the surface reveal little of the network of tunnels criss-crossing the interior. The pronotum of the adult beetle is a hooded structure which covers the top of the head down to the compound eyes.
SEM, ×60

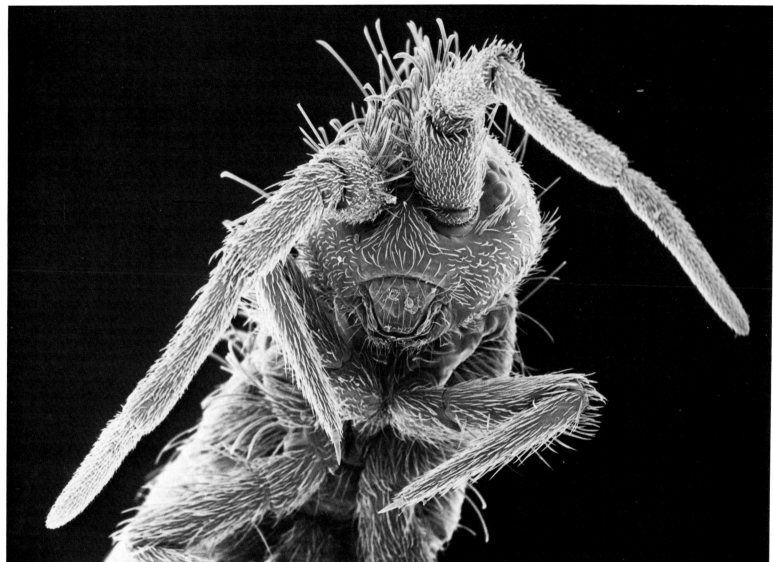

The portraits on these two pages represent a tiny sampling of the variety of structures found among insect heads. The images are all scanning electron micrographs.

3.26 The most primitive of all insects, dating back in the fossil record some 300 million years, is the springtail (family Collembola). There are about 1500 species distributed worldwide. The common name, springtail, is derived from their habit of using their tails to spring into the air. They have never developed wings and have no compound eyes. What resembles a compound eye, next to the base of the right-hand antenna of this springtail, is one of a pair of simple eyes, or ocelli, consisting of 6–8 light-sensitive cells. Below the floppy, segmented antennae are the biting mouthparts, sunk into the head capsule. Three legs, each

terminating in a claw, can be seen on the thorax.
SEM, ×40

3.27 The simple eyes or ocelli are clearly visible nestling in the forest of hairs behind the antennae of this species of springtail. The compact biting mouthparts are set well back under the front of the head. Some of the hairs covering an insect serve as sensory organs and are known as setae. These hairs are connected at their base to nerve cells which are stimulated by touch or by chemical means.
SEM, ×175

3.28 This nymph of the treehopper (family Membranacidae) will undergo further moulting before reaching adulthood. The seam visible along the top of the head splits and the new stage emerges. The nymph is covered in spiny projections remarkably similar

to those on the surface of the leaf on which it feeds. The proboscis, extending downward into the leaf, consists of an outer sheath which encloses slender piercing and sucking stylets used to drink plant juices. The short antennae are just visible beneath the compound eyes, which stand out like thimbles on either side of the head.
SEM, ×40

3.29 Distributed in clusters and repetitive patterns, the hairs on the moth fly *Psychoda sp.* present an exquisite appearance. The overall wooly effect produced by the hairs is the point of similarity between the fly and its namesake, the moth (see Figure 3.30). The compound eyes of this fly stretch like crescent moons around the sides of its face, almost meeting at top and bottom. The large, feathery antennae emerge from between the

crescent eyes. A pair of palps, or tactile feelers, are seen at the centre bottom of the head.
SEM, ×300

3.30 Scales are a common feature of butterflies and moths, which belong to the order Lepidoptera, meaning 'scale wings'. The diamond-backed moth *Plutella xyhostella* is covered in tiny overlapping scales. Scales are modified hairs, some broad and flat, some long and narrow, which lock into sockets on the body and wing membrane. The antennae, which sweep back and out of the picture, are positioned above the circular compound eyes. One of the moth's pair of palps is seen sticking upwards from beneath the eyes. The round tip of its sucking proboscis is visible beneath the palp.
SEM, ×70

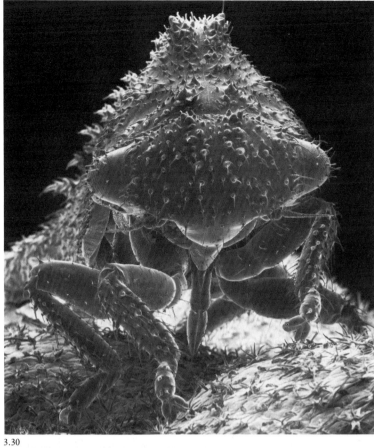

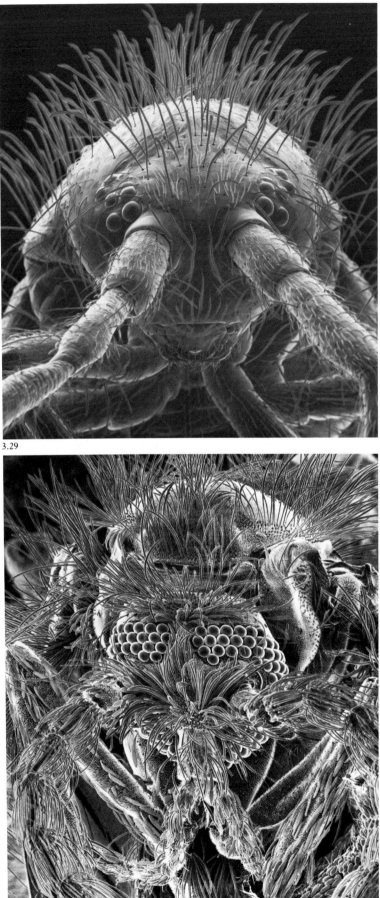

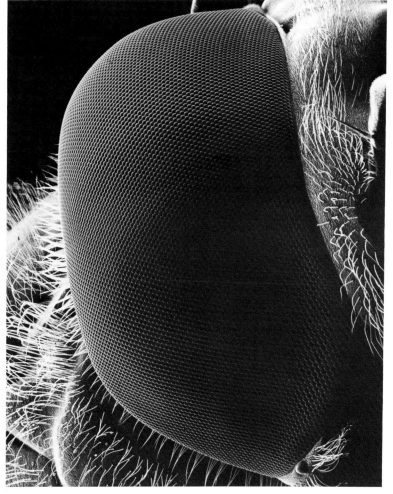

COMPOUND EYES

Sight is an important sense among most insects, aiding them in navigation, location of food and avoidance of danger. The size of the compound eyes varies enormously from a few hundred facets in some species to the 30 000 of the dragonfly. Adult insects that live mostly on the wing, such as bees and dragonflies, have more facets than ground-based insects such as beetles. The larger eyes provide a broader field of vision and the ability to rapidly detect prey and predators while in flight. Insects have colour vision, seeing blue, yellow and ultraviolet. Except for butterflies, few species see red.

The compound eye is composed of tightly packed light-sensitive units called ommatidia. Each is cone-shaped, with the lens at the wider surface and the light-sensitive cells tapering inward. The lens consists of a transparent cuticle and a crystalline cone immediately behind it. Light is focused from a narrow band of the external world onto a ring of cells behind the lens. The innermost margins of this ring are fused to form a light-sensitive rod. Light stimulating the rod is converted into electrical signals which are passed to nerves at the base of the ommatidium and from there to the brain.

In most compound eyes the light entering from one ommatidium is kept separate from that entering another by a layer of pigment cells which absorb any interference between the facets. The insect sees a mosaic image of the world made up from many separate segments. The image is poorly focused, but quick at detecting slight movement. In some nocturnal insects the pigment cells can be withdrawn at night so that light from several ommatidia is focused on one light-sensitive rod. In poor lighting conditions, this enables the insect to collect more light and produce a brighter image.

3.31 The eye of the honeybee *Apis mellifera* is adorned with long bristles, partly obscuring the hexagonal shape of the underlying ommatidia. Hairs on the body of the bee are associated with its role as a plant pollinator. The eyes too can perform this role. A pollen grain is just visible caught in the bristles at top right of the image. The base of one of the bee's antennae is at bottom right.
SEM, ×90

3.32 The large hemispherical eye of the hover fly *Syrphus ribesii* is smooth and contains many more facets than the eye of the honeybee in Figure 3.31. Each facet is positioned at a slightly different angle to its neighbour, thereby providing a slightly different view of the world. Compared to the eye, the antenna is tiny. It is visible at the top right edge of the picture as a small pad with a short bristle extension.
SEM, ×70

3.33 The dragonfly is a strong, swift flier that hunts, feeds and mates on the wing. Its compound eyes have the largest number of ommatidia among insects, equipping it with quick and precise vision. A few of its 30 000 ommatidia are seen in this light micrograph.
LM, filtered vertical illumination, ×480

CATERPILLAR HATCHERY

Most insects lay eggs. The eggshell, like the adult insect's exoskeleton, is made of a tough polysaccharide called chitin. The hatching larvae break through the shell either by chewing it or by muscular action. Bristles known as hatching spines may also assist in the procedure. The following scanning electron micrographs show the hatching of a larva, or caterpillar, of the large white butterfly *Pieris brassicae*.

3.34 The large white butterfly cements its eggs, like rows of tiny skittles, on the undersurface of a leaf, in this case a nasturtium. The larvae chew their way out of the eggshells, which provide their first meal. This may be a nutritional requirement, without which future development is impaired. In this picture two caterpillars are eating their way into the world, while a third has already emerged and departed.
SEM, ×110

3.35 The larva is protected from the attentions of predators by the fearsome bristles covering its body. The head capsule, from this angle, appears smooth except for the five buttons forming one of the caterpillar's simple eyes. Visible on the side of the first segment, like a small mound with a hole in it, is a breathing pore or spiracle, part of the respiratory apparatus. Air is taken up through such pores and carried around the body by a network of tracheal tubes. It diffuses directly from these tubes into each cell of the body.
SEM, ×100

3.36 Hatching takes a few minutes. The freed caterpillar immediately spins a safety net of fine silk threads which act as a lifeline back to the leaf should it fall. Strands of silk are seen wound around the leaf hairs and one thread is attached to one of the caterpillar's three pairs of true legs. The silk is produced by a pair of modified salivary glands.
SEM, ×50

3.37 The caterpillar feeds on the outer, waxy layer or cuticle of the leaf until the first moult, after which it eats the whole leaf. It chews the leaf by means of toothed mandibles, which are placed on either side of the lower face. Infestations of *P. brassicae* caterpillars can cause serious defoliation, particularly of cabbage plants, on which they are commonly found. The first three segments of the caterpillar each carries a pair of true, walking legs, one on each side of the body. The spiracles are noticeable along the side of the body.
SEM, ×90

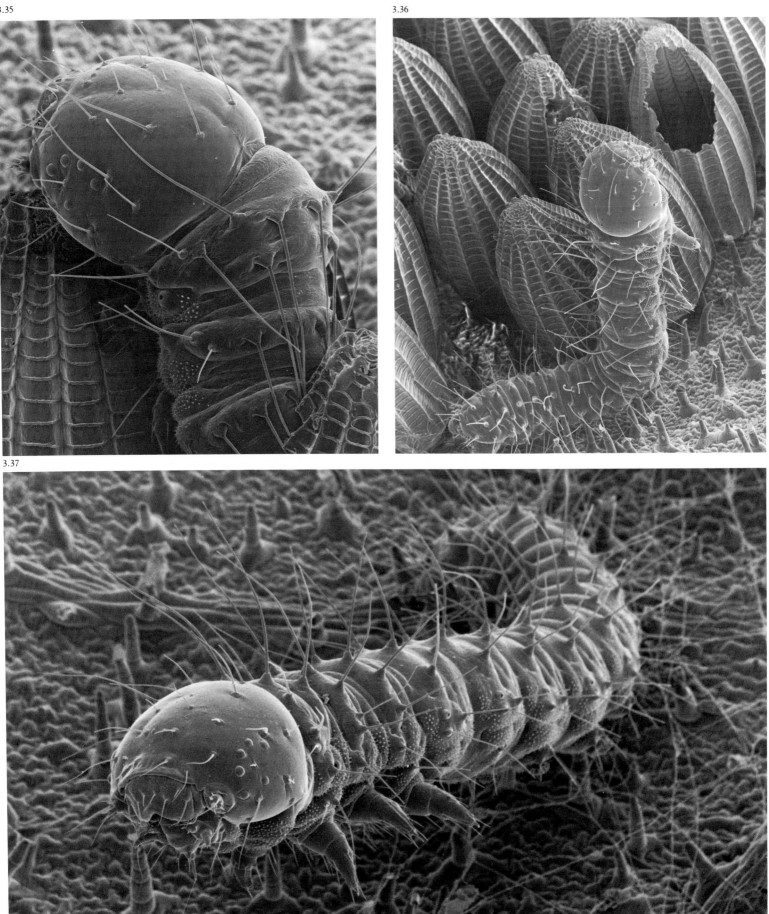

MITES

The mites belong to the Arachnida, which include the spiders and scorpions. There are some 15 000 known species of mites, spread throughout the world in a wide variety of habitats. Many are plant and animal parasites and some carry diseases – factors which make them of economic and medical importance.

3.38 *Glycophagus sp.,* the dust mite, is a common domestic nuisance. It thrives in damp houses, finding an abundance of food in the form of groceries, furniture stuffing, wallpaper paste and dust. This specimen was found in the dust taken from a household vacuum cleaner. It is seen among skin scales, fibres and cat fur.
SEM, ×625

3.39 The spider mites of the family Tetranychidae are major plant parasites and economic pests. They flourish on vegetable crops, fruit trees and flowers, sucking the fluid from leaf cells and causing the leaf to mottle, turn yellow and fall off. Plants are quickly infested with the mites. In this scanning electron micrograph, an unidentified spider mite is seen on the surface of a cannabis leaf. The legs are segmented, allowing for good flexibility in movement.
SEM, ×280.

3.40 Most mites are terrestrial, but some like this bright red, adult water mite *Hydrarachna sp.*, live an aquatic existence. The adults are free-living, although the larvae are parasites of some freshwater invertebrates, particularly water scorpions and water stick insects. The four pairs of segmented legs of the adult are seen here in the extended swimming position. The legs are used like oars to row the mite through the water. The long swim hairs visible on the rear two pairs of legs aid the process.
LM, Rheinberg illumination, ×90

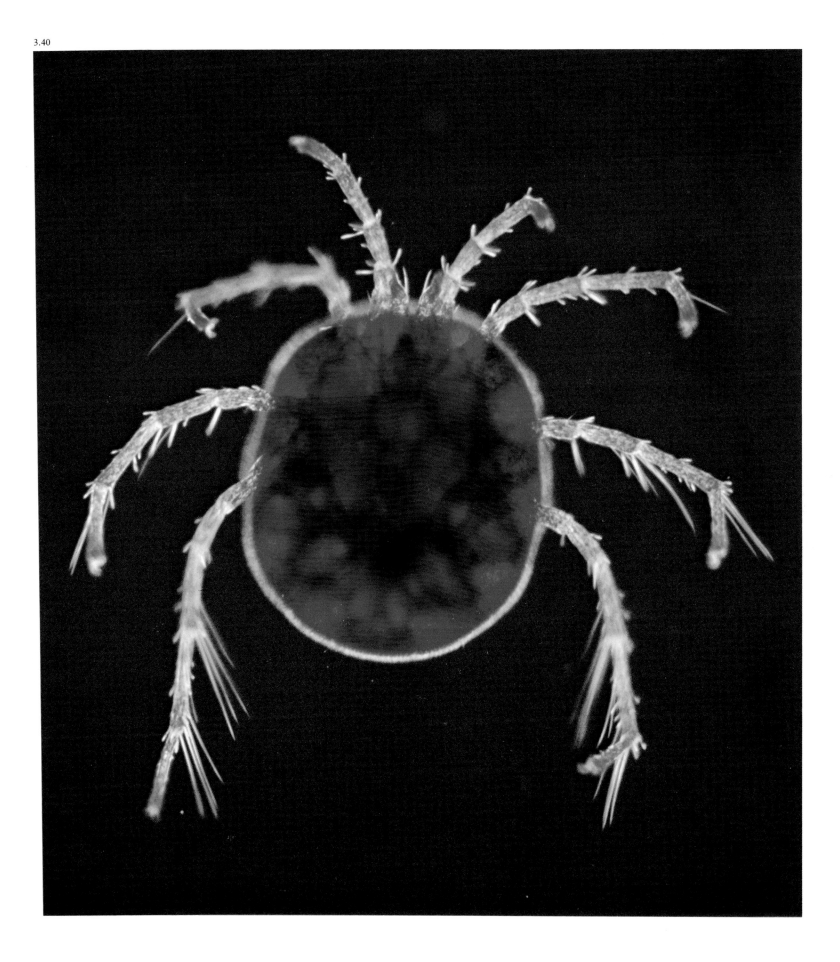

CHAPTER 4
SEED PLANTS

ACCORDING to modern science, life on Earth began about 3000 million years ago, in the warm waters of primeval oceans. Evolution has since produced the amazing diversity of creatures on Earth today. Early plant life consisted of simple organisms which lived and reproduced in water. Gradually plants invaded the land, but they remained dependent on external water for their successful reproduction.

About 350 million years ago, two inventions appeared which freed the land plants from their need for a watery or humid environment. One was the pollen grain. The other was the seed. The inventors were a group known as Gymnosperms. Today they are represented by conifer trees, cycads and the Japanese maidenhair tree, *Gingko biloba*.

The next evolutionary step took place about 130 million years ago. This was the appearance of the flowering plants, or Angiosperms, which dominate the plant kingdom today. Charles Darwin described the relatively sudden appearance of the flowering plants as 'an abominable mystery'. It was a remarkably successful event. There are about 250 000 species of flowering plants. They range in size from a tiny duckweed, *Wolffia arrhiza*, only about 1 millimetre across, to huge trees such as the coast redwood, *Sequoia sempervirens*, the tallest of which may reach 110 metres.

The success of the seed plants lies in their methods of reproduction and dispersal of their offspring. The male sex cells are produced within a tough, drought-resistant package, the pollen grain. This protects the sex cells during their initial journey to the female. After successful fertilisation, the offspring are packaged as seeds. These are dry structures, capable of surviving for long periods and travelling great distances without the need for water. Using these twin strategies, seed plants have colonised almost the entire land surface of the Earth.

The word 'flower' evokes a particular vision, of something coloured, possibly scented, beautiful – an adornment for our gardens and homes. Flowers corresponding to this vision in fact represent only a part of the wide spectrum of forms produced by evolution. Though largely unnoticed, flowers are produced in huge numbers by grasses, cereal crops and forest trees. The reason why some flowers are conspicuous and others are not is related to the method they use to achieve the successful union of male and female sex cells.

Inconspicuous flowers generally rely on the wind to carry their pollen grains for them. The wind cannot see, so there is no need to produce a showy flower. Coloured and scented flowers are designed to appeal to some living creature which will act as the bearer of the pollen. The most common pollinators are insects, but some plants use birds, bats, or even snails. They are drawn to the flower by sight or scent, and rewarded for their visit by gifts of nectar, or of some of the highly nutritious pollen itself. In specialised cases, the rewards are of a different nature. Many orchids, for example, pose as females of insect species. They are visited by hopeful males, who pollinate the flower while attempting to satisfy their own sexual desires.

The problems of living on dry land go beyond reproductive strategy. Life as a large land plant is altogether different from that as a small aquatic organism, let alone a single cell. Water is still needed for growth: it has to be extracted from the ground and conveyed to all parts of the plant body. The land plant therefore has to develop an extensive root system and an internal vascular network. In order to obtain energy for growth, the plant must expose green photosynthetic tissue to sunlight. For aquatic plants, this presents no mechanical problems – the water bears the weight of the plant tissues. On land, a strengthened stem has to be made to support the leaves. Exposed to the sun and the wind, the leaves themselves must be designed to reduce the loss of precious water to a minimum, while at the same time carrying on their photosynthetic function.

Throughout their evolution, plants have had to face predation from other living creatures. This hazard is particularly acute on land, where plants are prey to the attention of flying insects and birds, as well as animals.

Plants have solved these problems in a variety of ways. The 250 000 species of flowering plants could be said to represent 250 000 different solutions. It is only by the use of microscopy that the beauty and subtlety of some of them can be appreciated. The scanning electron microscope, in particular, has been a source of revelation to botanists. Many of the scanning electron micrographs in this chapter were produced by viewing the specimen in an intact frozen state. They show details of surface structure and function which could not have been revealed by any other method. Before the invention of the scanning electron microscope in the late 1940s, such pictures were unthinkable.

4.1 The first stage in the reproduction of a seed plant is the safe arrival of the pollen at the receptive female surface, called the stigma. In this scanning electron micrograph of a frozen flower of the turnip, *Brassica campestris*, a group of pollen grains is seen covering the stigmatic surface. Each pollen grain is egg-shaped, with a patterned surface and a furrow along its length. Two smooth projections from the female stigma are also visible. The pollen grain in the centre of the picture has produced a narrow pollen tube. This has penetrated the female surface. It will grow down into the flower until it reaches the vicinity of the egg cell. Inside the pollen tube, two sperms follow its growth. When the egg is reached, the tube bursts and releases the sperms, one of which will fertilise the egg. The whole process takes place without the need for external water.

SEM, false colour, ×350

ROOTS

Roots have two functions: they anchor the plant and extract water from the soil. Seeking water, they can grow to considerable depths. The roots of maize extend 1.5 metres down, but the roots of trees penetrate to 6 metres or more in light soils. The main root produces many branches, called laterals. Roots also produce fine hairs which serve to increase enormously the area of their absorbing surface. In a study of a four month old rye plant, it was estimated that the root system produced 14 billion root hairs and had a total surface area of over 600 square metres. This was 130 times greater than the surface area of the plant above ground.

In the tip of the root, special cells called statocytes are produced which can sense gravity and cause downward growth. They contain large and dense starch grains which act as gravity detectors. Later in life, these same cells are pushed outwards to the surface of the root tip, where they produce a slimy mucus to lubricate the passage of the root through the soil.

Water absorbed by the root hairs is transferred to the vascular system of the plant. In roots this system consists of a central cylinder of tissue containing *xylem* and *phloem* cells. The xylem cells carry water from the roots to the rest of the plant. The phloem cells bring supplies of sugar from the leaves to sustain the growth of the root tip.

4.2 Root hairs are short-lived, delicate structures. Roots grow from the tip, and new hairs are continually produced from a region just behind it. The process is clearly visible in this very low-magnification scanning electron micrograph of a frozen cabbage seedling. Near the root tip, the hairs are young and short; further

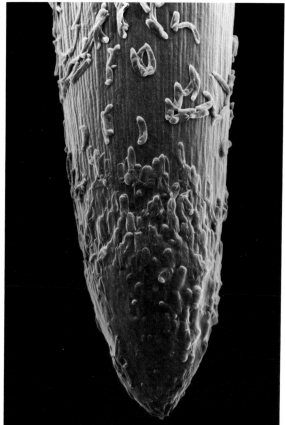

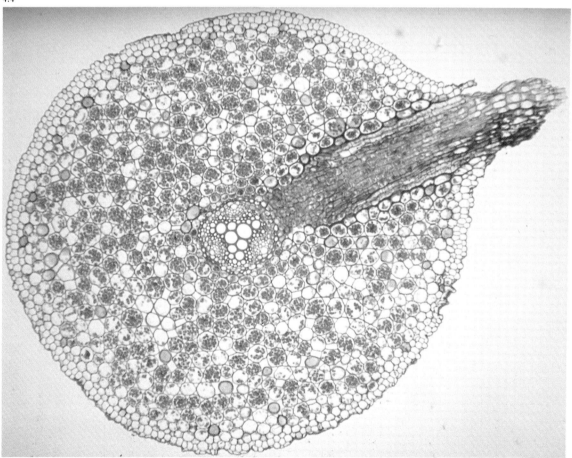

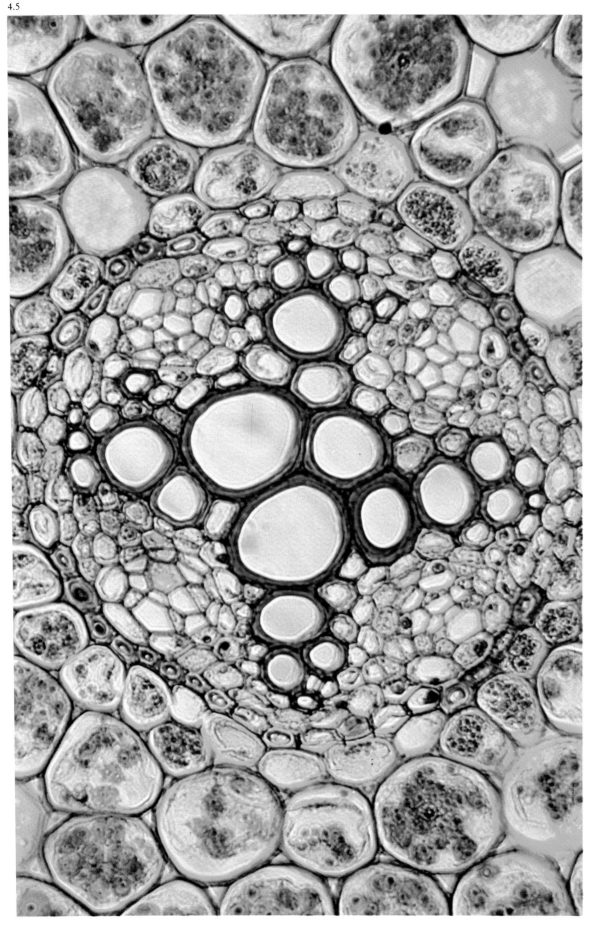

away they are longer. The picture also shows how the coat around the seed at the top right has split open to allow the root to emerge. Inside the split seed, the first leaves have yet to unfold.
SEM, ×12

4.3 The root tip has to push its way through the soil. To assist this operation, it continually produces a loose layer of slimy cells at its outer surface. In this picture of a laboratory-grown root of wheat, these cells can be seen embedded in the slime layer, which appears as a smooth covering on the frozen specimen. During growth in soil, the cells are rubbed off and die.
SEM, ×60

4.4 Laterals are produced deep within the main root from a special layer of cells surrounding the vascular tissue, called the pericycle. This light micrograph shows a developing lateral in the root of the buttercup, *Ranunculus acris*. As the young root lateral pushes outwards, it crushes and kills the cells of the root cortex which are in its way. The picture shows that many light microscope stains are not very specific in their action. The green dye has stained cell walls only, but the red dye has coloured starch grains (small particles within the cortical cells) as well as the thickened walls of the xylem cells in the centre of the root and other cells in the lateral itself.
LM, bright field illumination, stained section, ×22

4.5 In this close-up of the central region of a root similar to the one in Figure 4.4, the details of the vascular region, or stele, are revealed more clearly. In the centre, arranged in the form of a cross, are the large empty xylem cells, their walls stained red. Between the arms of the cross at the edge of the circular stele region are four groups of other cells, also red. These are the phloem. Beyond the stele lies the cortex of the root with its large cells containing red-stained starch particles.
LM, bright field illumination, stained section, ×870

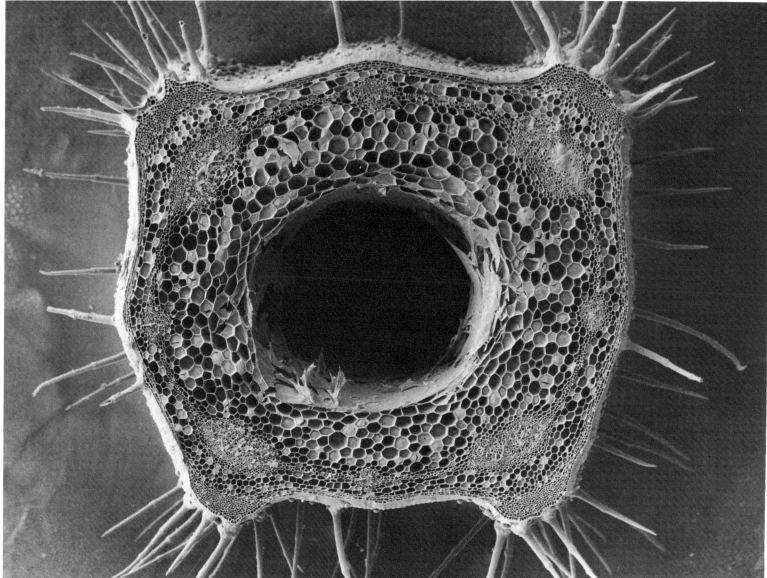

STEM & WOOD

The stem of a plant connects the leaves with the root system. Water and soil minerals travel up the stem in the cells of the xylem, and sugar solution flows down the stem in the phloem. The stem has an important additional function: it supports the entire weight of the aerial parts of the plant. This weight may be slight in the case of a small annual plant. A long-lived tree, however, may present a formidable problem. The most

massive living thing on Earth is a tree. The giant sequoia known as 'General Sherman' – a specimen of *Sequoiadendron giganteum* – is estimated to weigh over 2000 tonnes.

Stem growth is classified into two types. Primary growth occurs in annual plants. The stem tissues consist of vascular cells, together with unspecialised cortical cells. Strength is gained by the formation of groups of cells called collenchyma, which have very thick walls. Primary stems are often hollow, because a hollow tube resists bending better than a solid rod of the same weight.

Secondary growth occurs in long-lived perennial plants with

persistent stems, of which the most obvious examples are trees. Each season, the stem grows thicker by the action of a ring of soft tissue, the cambium, which is located under the bark. The cells produced towards the outside of the cambium become phloem; those towards the centre of the stem become xylem. The xylem cells develop massively thickened walls made of cellulose together with an inert filler called lignin. Once this thickening has occurred, the cells die, remaining as a system of extremely strong and resilient tubes for the passage of water. This is the tissue which we call wood.

4.6 The scanning electron micrograph of the cut stem of the white dead nettle, *Lamium album*, shows all the characteristics of primary growth. The hollow centre is produced by the collapse of thin-walled pith cells. Surrounding this core is a layer of unspecialised cortical cells – the cortical parenchyma. Eight 'vascular bundles' serve to conduct water and nutrients: one near each of the four corners, and one half way along each side of the stem. In the extreme corners of the stem – the most advantageous position mechanically – small collenchyma cells are visible. The spikes on the outside of the stem are hairs, designed to discourage crawling insects from climbing the plant.
SEM, ×30

4.7 Trees grow in annual cycles, and this gives rise to the familiar growth rings visible on sawn logs. In this radial section of wood from the conifer *Tsuga canadensis*, the hemlock tree, the cycle is shown in terms of the width of the cells comprising each year's growth. The four bright red bands running diagonally from top right to bottom left correspond to four years of growth in the tree's life. Early in the season, the cells are widely spaced, but they bunch closer together as autumn approaches. Thus spring of one year is at the very top left of the picture; the bright red lines are widely spaced until, as autumn comes, they bunch together to form the thick red band; then, in the following spring, the pattern is repeated. The darker red bands running at right angles to the annual rings are 'rays', a system of horizontal xylem cells. The bright colouring of the micrograph results from the interaction of polarised light with the ordered cell wall structure of the xylem.
LM, polarised light, ×660

4.8 The structure of hardwoods is more complicated than that of the softwoods produced by conifers. There are more different types of cells and, in particular, hardwoods have very large water-conducting elements called vessels. This cross-section of hardwood from the stem of the sugar maple tree, *Acer saccharum*, was produced by the same polarised light technique as the previous picture. The large blue cells are the vessels. The red bands are cells of the horizontal xylem system, the rays, which distribute water to the cambium and also act as storage centres for starch and lipids. There are no annual growth rings visible in this section.
LM, polarised light, magnification unknown

LEAVES

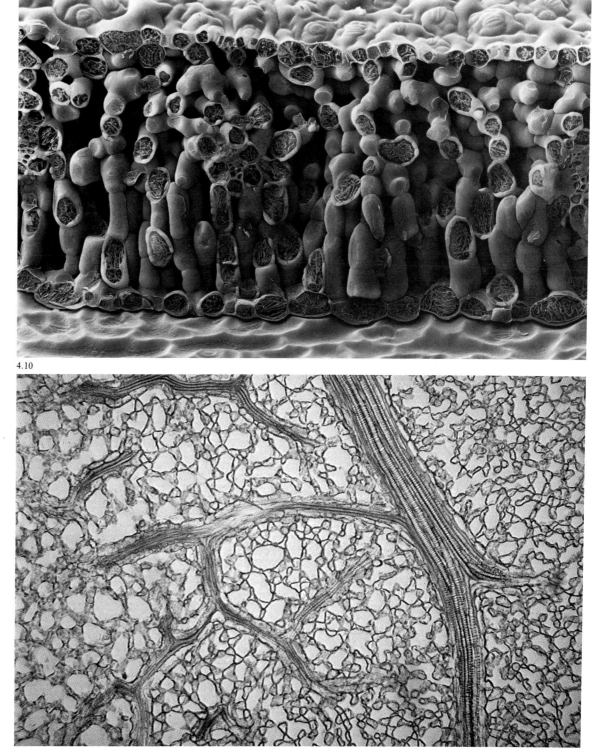

4.10

Leaves are the site of photosynthesis – the conversion of atmospheric carbon dioxide into sugar. This chemical process provides the plant with energy and, as a waste product, produces the oxygen we breathe. The chemistry takes place inside leaf cells, in organelles called chloroplasts, which are described on pages 116–117 in the chapter on the cell.

The raw materials of photosynthesis are sunlight, water and air. The first two present little problem. The typical leaf is a flat blade designed to catch the light, and its water supply comes via the stem from the roots. The difficult supply problem concerns the air. A leaf exposes a large surface area to the atmosphere and, because of this, it has to be covered with a waterproof layer called the cuticle. Without it, the plant would lose water too rapidly, and die. But the cuticle also stops air getting into the leaf, so it has to be perforated with a series of pores. These pores, or *stomata*, are held open during the day, and are closed at night in order to conserve water.

The surface of a leaf is rarely a flat landscape when seen through a microscope. A wide variety of specialised features comes into view. These outgrowths are collectively known as trichomes, and scanning electron microscopy reveals their diversity particularly well, as Figures 4.14–4.18 show.

4.9 A leaf is built like a sandwich. To see inside it with a scanning electron microscope, the leaf must first be frozen and broken open. This picture is of a leaf of the turnip, *Brassica campestris*, prepared in this way. The single horizontal lines of cells towards the top and bottom of the picture form the skin, or epidermis, of the leaf, covered with the cuticle layer. Special cells in the epidermis control the size of the stomatal pores, most of which occur on the underside of the leaf, which is at the top of this picture. The loosely packed cells in the interior of the leaf are called mesophyll cells, and they are the site of most of the photosynthesis. Some are intact, others are broken open.
SEM, ×250

4.10 An entirely different view of a leaf is given by the light microscope. This section of the leaf of a privet, *Ligustrum vulgare*, is dominated by the leaf's vascular system, which is seen as a branching network. The mesophyll cells, loosely packed together between the vascular strands, are small and irregularly shaped; the large clear areas amongst them are air spaces.
LM, bright field illumination, ×180

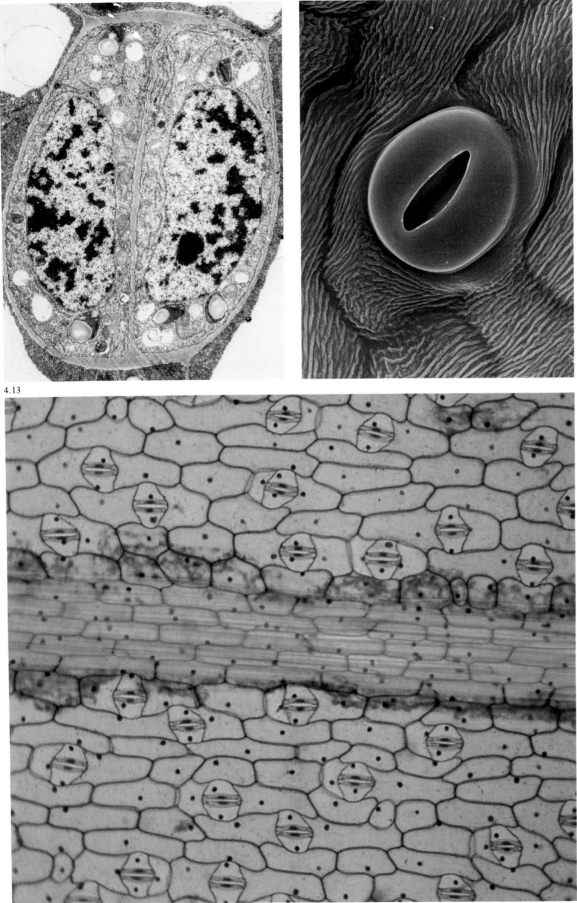

4.11 To see details within cells, a transmission electron microscope must be used. This pair of cells in the epidermis of a leaf of the garden pea, *Pisum sativum*, is at an early stage of forming a stomatal pore. Each cell contains a large centrally placed nucleus, oval in outline and containing genetic material stained black. Also visible are a few starch-containing chloroplasts – the five dark grey bodies in the cytoplasm, four of which are seen to have pale grey contents (the starch). The wall between the two cells will soon split down the middle. This will open an airway into the leaf's interior, which is effectively behind the page as it is viewed. The size of the pore will depend on the activity of the two cells, which are called guard cells. When they swell up, the pore will be pushed open; when they shrink, it will close.
TEM, stained section, ×5000

4.12 This stomatal pore on a sepal of *Primula malacoides* at first appears to be open. In fact, it is closed. The opening is a permanent gap in the cuticle; beneath it can be seen, tightly pressed together, the walls of the guard cells which control the pore. The ridges on the sepal surface are a peculiarity of this plant; the deeper furrows reveal the outline of the adjacent epidermal cells.
SEM, ×750

4.13 This surface section of a leaf of *Tradescantia* under the light microscope shows the distribution of stomata. The paired guard cells are stained red-brown; the small dark brown particles elsewhere are nuclei of epidermal cells. The broad blue band across the picture corresponds to a vascular strand, or leaf vein.
LM, ×445

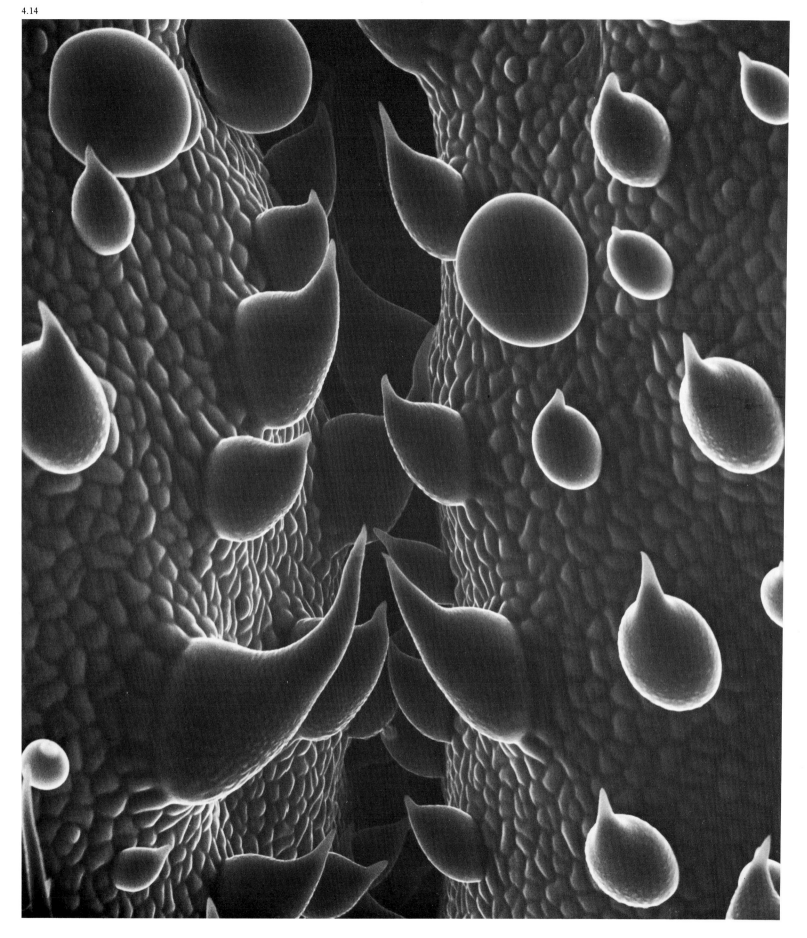

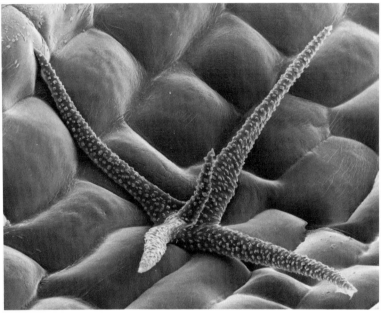

4.14 Flowering plants produce a wide range of substances designed to reduce their palatability to animals. The trichomes on this leaf of the marijuana plant, *Cannabis sativa*, are sites for the formation of two of these 'secondary products'. The pointed outgrowths are cystoliths, which contain tough crystals of the mineral calcium carbonate. The rounded structures towards the top of the picture are glandular trichomes containing tetrahydrocannabinol resin. It seems ironic that evolution has ensured the survival of this species by means of a poison which has proved highly attractive to human beings.
SEM, ×850

4.15 These branched trichomes completely obscure the surface of the leaf which produces them. Designed to reflect sunlight and keep the leaf cool, they belong to the plant *Verbascum pulverulentum*, popularly known as hoary mullein. The thick, hairy covering of trichomes was scraped from the leaves by neolithic flint miners at Grimes Graves in Norfolk. They are believed to have rolled it like cotton to use as wicks in oil lamps.
SEM, ×105

4.16 *Primula malacoides* is a popular pot plant which looks as though it has been dusted with flour. The reason is the production of wax from stalked trichomes, seen here on the surface of a sepal. The wax, called farina, insulates the plant and prevents condensation which would otherwise block the stomatal pores between the trichomes.
SEM, ×335

4.17 This branched trichome is not on the surface of a leaf, but inside it. The leaves of the water lily, *Nymphaea alba*, are permeated with wide passageways which supply air to the submerged roots via the stem. The internal surfaces of these airways are covered with hard trichomes encrusted with crystals of poisonous calcium oxalate. Their function is to deter inquisitive burrowing insects.
SEM, ×120

4.18 Plants have to conserve water, particularly in winter when it is difficult to obtain from cold or frozen ground. The Japanese evergreen shrub *Eleagnus pungens* employs these umbrella-like trichomes for this purpose. Found on the underside of the leaves, the trichomes overlap and completely protect the leaf from the effect of drying winds.
SEM, ×80

4.19

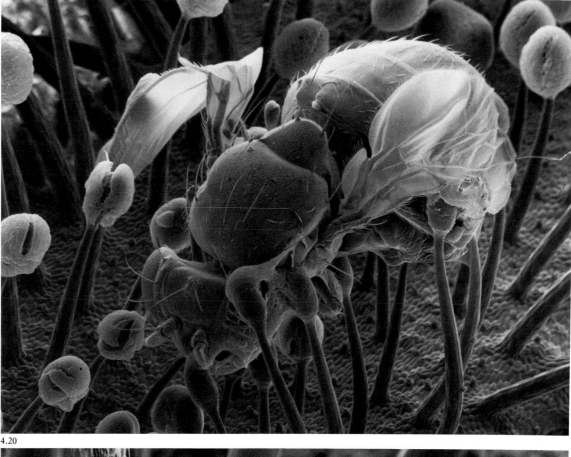

Plants have an ambivalent relationship with the animal kingdom. They often depend on insects or other animals to achieve pollination of their flowers and dispersal of their seeds. But they also represent nothing more than a free lunch to hordes of bugs, aphids and caterpillars, not to mention more formidable adversaries in the shape of herbivorous mammals. Plants guard against such predation by synthesising poisons, or by covering themselves with spines, sharp bristles, or sticky glues.

Some of these defence mechanisms are astonishingly subtle. All species of potato, for instance, respond to wounding by the rapid production of a chemical which inhibits the enzymes used in an insect's digestive system; in effect, they make themselves indigestible.

Such chemical defences do not always work as intended. Plants of the milkweed family contain a strong cardiac poison, fatal to vertebrates. Insects, which are unaffected by it, have learned to gain immunity from their own vertebrate predators by eating the tissues of the plants.

Plants are not always passive, defensive creatures. Some inhabitants of boglands are able to obtain the nitrogen and minerals which the soil lacks by capturing and digesting insects. There are about 500 species of these carnivorous plants, and they have developed a variety of methods for immobilising their meal. One example is the sundew.

4.20

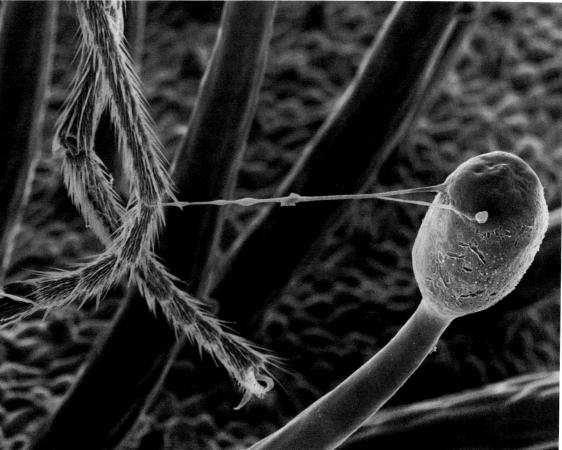

4.19 The upper surface of a sundew leaf is covered with stalked trichomes, each bearing at its tip a gland which secretes a strong adhesive. This scanning electron micrograph of a frozen leaf of the Cape sundew, *Drosera capensis*, shows the glands in

action. A small fly of the Psilidae family has wandered onto the leaf, and is stuck fast. The more the fly struggles, the more glands it touches, until eventually it becomes immobilised.
SEM, ×30

4.20 This detail of the same specimen as in Figure 4.19 shows that the initial contact between plant and insect may be very tenuous. Two of the fly's legs appear in the picture, one showing the pair of tiny claws which it uses for gripping rough surfaces. A very thin strand of adhesive attaches the other leg to the tip of a trichome. At this stage, the fly may escape if it is strong enough. But if it doesn't, its struggles will set up electrical signals in the trichome's stalk. Over a period of perhaps 30 minutes, these signals will cause the stalk to bend inwards towards the fly. Once in contact with the fly's body, the gland secrets a digestive juice and absorbs the nutrients which this releases from its prey.
SEM, ×125

4.21 The leaves and stem of the stinging nettle, *Urtica dioica*, are covered in hairs. The majority of them are of simple construction, but a few are modified into stings. Stings occur in groups of two or three, mostly along leaf veins. This picture shows both the base of a sting (the thick structure in the background) and the tip of another sting in the foreground. The wall of the sting is brittle and contains silica. At its tip, a small spherical cover of a glass-like material seals the end. The slightest touch to this seal causes it to break, uncovering a sharp hollow needle finer than a hypodermic. The injection which results is a cocktail of two chemicals, acetylcholine and histamine. The nettle's sting is an effective defence against grazing animals but largely useless against insects.
SEM, ×250

FLOWERS

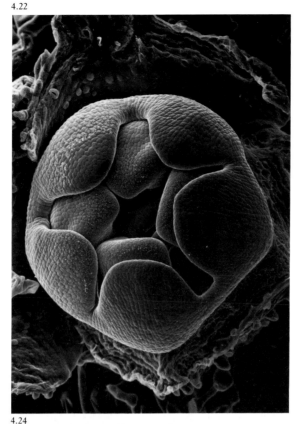

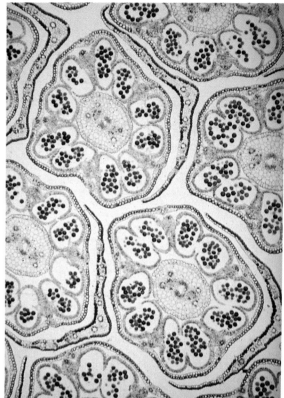

The bewildering diversity of form, colour and scent of flowers has one purpose – the reproduction of the plant. The idealised flower consists of a series of different structures arranged in concentric layers. On the outside there is a layer of protective sepals, then a ring of petals, then a ring of male organs called stamens. Each stamen consists of a stalk with an anther at its tip. The anther is the site of pollen production. In the centre of the flower is the female tissue, the carpel. This contains the egg cells, and it also develops a receptive surface called the stigma. In many flowers the stigmatic surface is produced at the end of a stalk growing from the top of the carpel. In any particular flower, each of these layers may be modified, or absent.

Flowers may be produced singly, or in complex groups comprising thousands of individuals, as in the daisy family. Some are microscopically small; on the other hand, a species of *Rafflesia* produces a bloom nearly 1 metre in diameter and weighing 7 kilograms. Nor are all flowers delicate and innocent; the African water lily *Nymphaea citrina* systematically drowns its insect visitors to wash pollen from their corpses onto its female parts.

4.22 A flower begins its life as a series of small bumps on the side of a shoot tip. Its final form is not generated until after considerable growth. This small bud of the snapdragon, *Antirrhinum majus*, appears almost symmetrical; the flower it produces is highly asymmetric. The sepals have been dissected away so that the outer, five-lobed layer in the picture represents the petals. Inside them, the tips of four developing anthers can be seen. In the centre of the bud, the cleft surface of the stigma has formed.
SEM, ×80

4.24

4.23 Each 'flower' of a daisy such as *Cosmos bipinnatus* is a mass of thousands of tiny floral units or florets. Those at the edge – the 'petals' – are sterile. This light micrograph shows a section through a group of fertile florets in the centre of the *Cosmos* inflorescence. Each floret contains five paired anthers, recognisable by the mass of brown-stained pollen grains within them. They surround the carpel tissue, stained pale blue; the developing ovules within the carpel are the small orange circles.
LM, bright field illumination, ×50

4.24 The flower of the chickweed, *Stellaria media*, conforms closely to the idealised form. This low-magnification scanning electron micrograph shows the five sepals on the outside, with five deeply lobed petals inside them. The three granular objects are anthers, covered in pollen grains. In the centre of the flower is the carpel, with three rough stigmatic surfaces visible.
SEM, ×15

4.25 This picture is of the same species as in Figure 4.24. The view is from the side, and was obtained by dissecting away part of the flower and tilting it in the microscope. It shows the bulbous carpel, with the stigmas emerging from it on the end of three short stalks, or 'styles'. Two of the long-stalked anthers have moved to make contact with the stigmas; the one on the right of the picture has remained separate and clearly visible, its tip covered with pollen. The smooth round objects at the base of two of the anther stalks are drops of nectar. This is a sugary substance offered to insects as a reward for visiting the flower and pollinating it. Curiously, a chickweed is quite capable of pollinating itself and often does so in the wild. The round pollen grains visible on the stigmatic surfaces of this flower show that it has been pollinated, probably by the two anthers in contact with the stigmas.
SEM, ×40

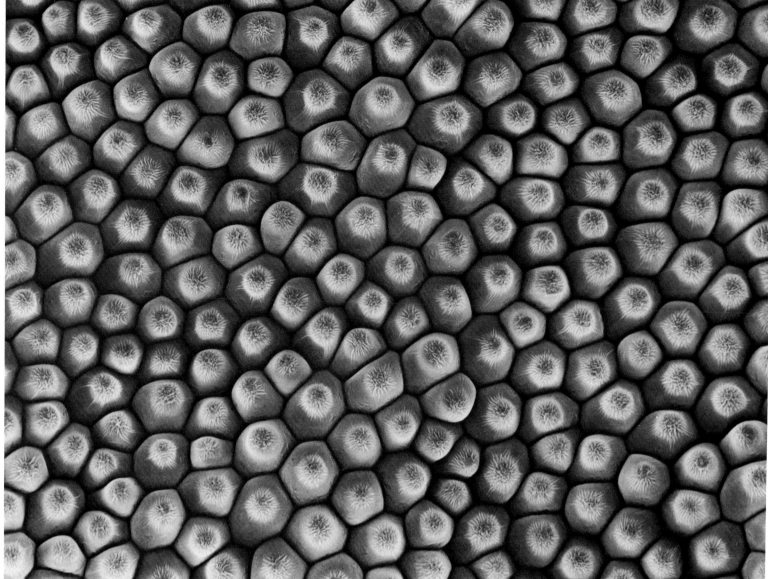

4.26 The petals of most flowers are in fact stamens which have switched roles during evolution from sexual organ to advertising agent. Although they appear smooth to the eye, they often reveal fine sculptured detail when viewed with a microscope. This rose petal surface consists of closely packed cells, each one of which is topped with fine ridges. The reason for this is unknown, but may be related to the reflection of light from the petal.
SEM, ×365

4.27 The stigma is the first point of contact between the pollen and the female organs which it is to fertilise. The stigmatic surface is usually moist and sticky, and often feathery or brush-like; this increases the area available for pollen to land on, and so improves the chances of a successful union. In this scanning electron micrograph of an *Hibiscus* flower, five separate stigmas are visible. Not every pollen grain which lands on a stigma is accepted by the female. Cross-pollination between different plants of the same species is the general rule in plant reproduction. The stigma will inhibit the growth of pollen which is not of the correct species. In many cases, it will also reject pollen from other flowers of the same plant.
SEM, ×20

4.28 Flowers are not all hermaphrodite. Some are exclusively male, others female. In such cases the tissues corresponding to the other sex are absent from the flower. This group of florets from the common daisy, *Bellis perennis*, are all female. The picture shows the developing stigmatic surfaces at the tips of a group of carpels. There are no stamens present. Interestingly, most of the flowers are producing five stigmas, but one has developed abnormally and has six.
SEM, ×35

4.29 The anther first develops as a closed chamber divided into four spaces. As the pollen grains develop, the wall between pairs of spaces is broken down, and at maturity the anther consists of two closed pollen sacs. This mature anther is from shepherd's purse, *Capsella bursa-pastoris*. The coarsely textured surfaces are the outside of the anther wall. The finer surface is the inside of the same wall. It is visible because the anther has split open to allow the pollen to escape – a process called dehiscence. A dozen or so pollen grains remain inside the opened pollen sac.
SEM, ×220

4.30 This picture is a close-up of one of the anthers of the chickweed specimen in Figure 4.24. In chickweed, the pollen sacs open very wide indeed, effectively turning themselves inside out. This results in the appearance shown here, of a mass of pollen grains lightly adhering to the inside wall of the sac. In wind-pollinated flowers, the grains are wafted away on air currents. Chickweed pollen simply waits. If an insect happens to pass, then grains may be brushed onto its body and effect a distant act of cross-pollination. If no insect appears, the anther stalk eventually bends inwards so that the pollen will fertilise its own flower.
SEM, ×150

4.27

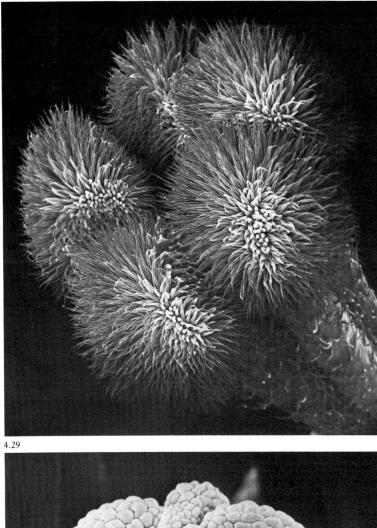

4.28

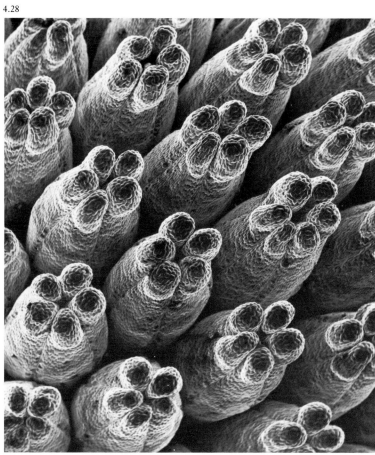

4.29

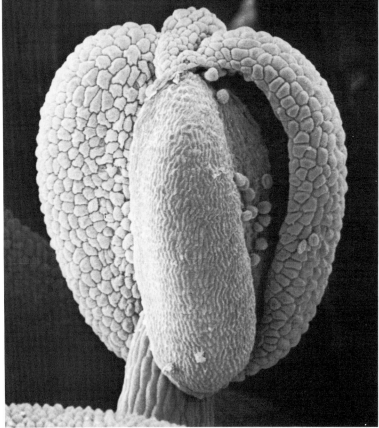

4.30

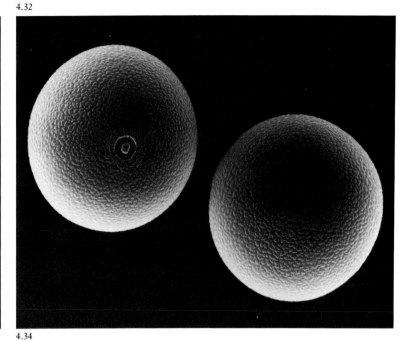

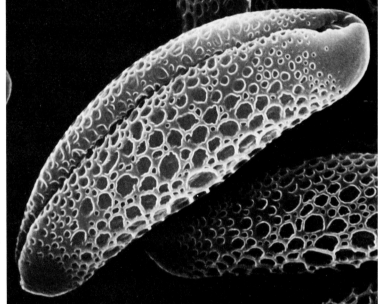

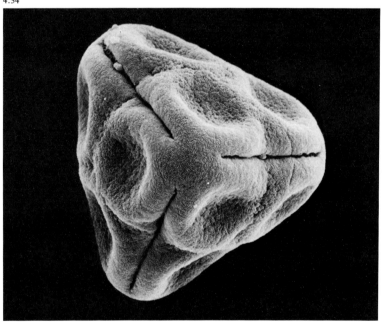

POLLINATION

All pollen looks much the same to the unaided eye – like fine yellow dust. Microscopy reveals that pollen grains differ widely in size, shape and surface texture. Those of the alpine forget-me-not, *Myosotis alpestris*, are only 3 micrometres across, whereas cucumber pollen (*Cucurbita pepo*)

has a diameter of 200 micrometres. About half the flowering plants produce ellipsoidal pollen grains, but spheres, polyhedra and long thin rods are also encountered. The surface varies from almost smooth to highly textured; small spikes, ridges and furrows appear in intricate patterns. Smooth pollen grains are found particularly in wind-pollinated flowers; they separate from each other easily in preparation for flight. Textured grains are designed to adhere together and to the hairs of pollinating insects.

The surface pattern on a pollen grain is in some cases characteristic enough to allow the species to be identified. The wall contains a very inert polymer called sporopollenin, as a result of which the pattern may be preserved for hundreds or thousands of years. This is true of pollen preserved in peat bogs, and is of archaeological value.

The task of the pollen grain is to carry the male sex cells to the female stigma. If the partners are compatible, the pollen grain germinates to produce a thin tube

– the pollen tube. This grows through the stigmatic surface and down to the ovule, where it bursts, releasing two sperms.

Compatibility depends on the recognition of proteins released by the pollen. These can be highly allergenic to humans. Hay fever results from breathing in pollen grains of wind-pollinated flowers, especially grasses. The amount of pollen produced by such flowers can be enormous; a single catkin of the hazel, for example, produces about four million pollen grains.

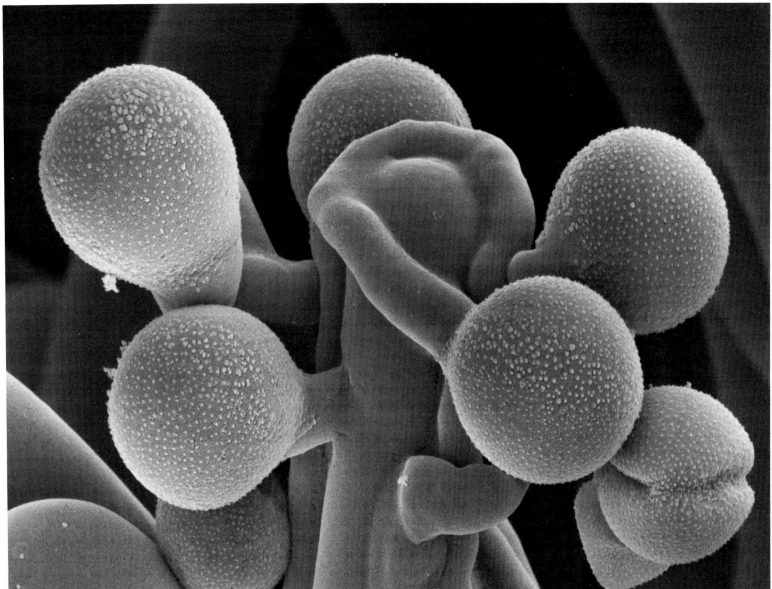

4.31 Pollen grains are a favourite specimen for scanning electron microscopy. Not only beautiful, they are often tough enough to need no chemical fixation or freezing. The single grain in this picture is from the passion flower, *Passiflora caerulea*. Passion flowers provide copious nectar, and are insect-pollinated. A newly opened passion flower has its anthers fully exposed, but no female parts. Over the course of a few days, it unloads its pollen onto passing insects, then the anthers shrivel. Their place is taken by newly developed stigmas on the ends of elongated styles. This separation of the sexes guarantees cross-pollination. The pollen tube emerges from one of the three curved furrows visible here.
SEM, ×1275

4.32 This spherical pollen is typical of many grass species, which are wind-pollinated. It is from the cocksfoot grass, *Dactylis glomerata*, a plant responsible for much early summer hay fever in Europe and North America (where it is known as orchard grass). The allergenic proteins are localised in the germination pore of the pollen wall. The grain on the right has its pore visible.
SEM, ×1065

4.33 The pollen of *Billbergia nutans*, a Brazilian member of the pineapple family, has a deep furrow along its length. The pollen tube is produced at one end of this furrow.
SEM, ×1475

4.34 This particle is not a single pollen grain, but four grains arranged in the form of a tetrahedron. All pollen grains initially develop in groups of four, called tetrads. In most plants subsequent growth results in their separation. In some species, such as this heather, *Erica carnea*, separation does not occur. Three of the four members of the tetrad can be seen. Each has three shallow depressions on its surface, and is bounded by a deep furrow through which the pollen tube or tubes will emerge.
SEM, ×1900

4.35 This view of germinating pollen grains of the opium poppy, *Papaver somniferum*, was obtained by freezing the flower before putting it into the scanning electron microscope. Eight pollen grains appear in the picture, clustered round a single finger-like projection, part of the stigmatic surface. Several pollen tubes are visible, most notably the one which encircles the top of the stigma before eventually growing downwards towards the base of the flower. The pollen grain at bottom right has not germinated; it retains the surface furrows which are characteristic of the species. Once germination occurs, the grain swells and the furrows disappear. At least five of the pollen grains in the picture have germinated, and this will result in the formation of a corresponding number of seeds. These develop from the ovules within the carpel, and as this occurs, the carpel tissue itself expands to produce the familiar seed capsule of the opium poppy. A single capsule may contain 2000 seeds.
SEM, ×1080

EMBRYO

Pollination and the growth of the pollen tube result in the release of two sperms into the embryo sac, the part of the ovule which contains the egg. One sperm fertilises the egg itself. This produces a cell called a zygote, which will eventually grow into the embryo. The other sperm fuses with two nuclei to form a special tissue called endosperm. The endosperm grows quickly, and serves as a source of nutrition for the young embryo.

These events all occur within an ovule. A flower may contain many ovules, each capable of producing one seed if it is fertilised. Apart from the embryo sac, the ovule consists of a nutritional tissue called the nucellus, together with a covering, or integument.

The development of plant embryos follows a variety of patterns, depending upon the species. In broad outline, the zygote divides and initially produces a short filament of cells, known as the suspensor. One end of the suspensor is anchored to the embryo sac, while the other end grows into the endosperm tissue. At this free end, the terminal cell divides repeatedly to produce a compact mass of small cells which are initially all the same in appearance. Once this globular mass reaches a certain number of cells, usually a few hundred, specialisation of the tissue begins to occur. The shoot tip, root tip, vascular system and first elementary leaves are formed. After this, the embryo grows rapidly, laying down storage materials such as starch and protein to sustain it during the initial stages of its eventual germination. The food supply for the growing embryo comes from the tissues around it, but also from the parent plant through the stalk connecting the ovule to the carpel wall.

The final stage in embryo development involves loss of water. The embryo shrinks in size and the connections between it and its parent are severed. The ovule develops a hard protective covering from the integument layers. Finally the embryo is dry, full of food reserves, and protected by a tough coat. It has become a seed.

4.36 Conifers belong to the group of plants called Gymnosperms, which means 'naked seeds'. This refers to the fact that the ovules, and hence the seeds, are carried on the outside of the female flower, not enclosed in a carpel. In this light micrograph of a female pine cone, the ovules appear as egg-shaped objects around the core of the cone. They are accessible to wind-borne pollen which falls between the covering of scales.
LM, magnification unknown

4.37 The other group of seed plants is the Angiosperms, meaning 'enclosed seeds'. To obtain this view of the ovules of the opium poppy, *Papaver somniferum*, the developing seed capsule had to be cut open. The picture shows the ovules attached to the mother plant by means of a short stalk, the funiculus. Each ovule will eventually become a seed.
SEM, ×45

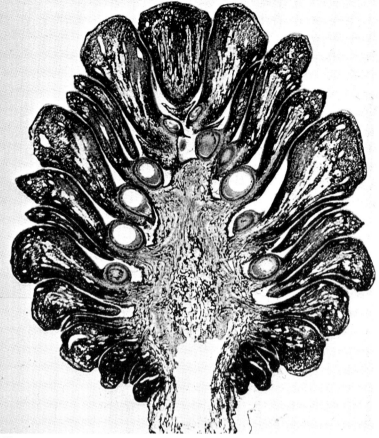

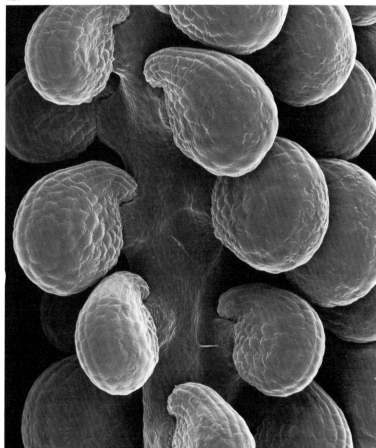

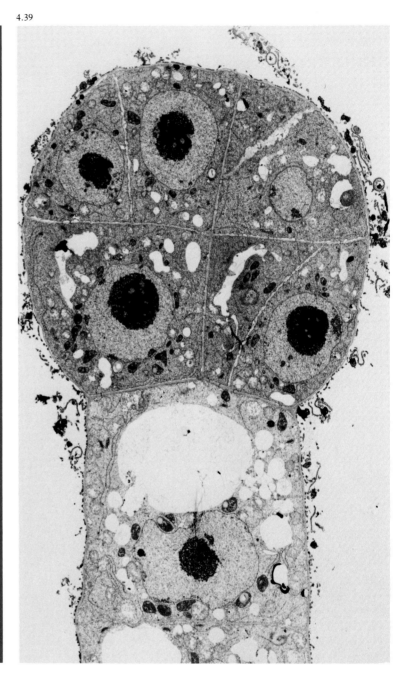

4.38 This embryo of the turnip, *Brassica campestris*, has been dissected out of the ovule two days after pollination took place. It consists of ten cells: six in the elongated suspensor, and four in the small globular tip at the top. The material at the bottom of the suspensor is remnants of the embryo sac. The cells of the globular tip will go on to produce the embryonic shoot and leaves, while the root will be formed from the suspensor cell nearest to the globular tip. The colours result from use of polarised light in the Nomarski Differential Interference Contrast microscope.

LM, Nomarski DIC, ×385

4.39 This transmission electron micrograph shows the globular tip and the first suspensor cell of an embryo like the one in Figure 4.38, but after one day's additional growth. The globular tip now contains sixteen cells, although this section, which is just 70 nanometres thick, only includes seven of them. The large grey bodies with black centres which can be seen in most of the cells are their nuclei. The large white areas in the suspensor are water-filled spaces called vacuoles. The material adhering to the outside of the embryo is fragments of endosperm tissue.

TEM, stained section, ×2600

SEEDS

Seeds have two purposes. They enable plants to survive through cold or drought – without seeds there would be no annual plants in temperate or arctic regions. And they make it possible for plants to colonise new habitats.

The dispersal of seeds involves a variety of mechanisms. Some seeds are simply very small and light, and are carried by air currents. The winter-flowering succulent *Kalanchoe blossfeldiana* has seeds which weigh only 1/100 000th of a gram. Others develop wings or propellers and can fly, sometimes for great distances. Some attach themselves to passing animals by stealth, but others are presented within colourful, tasty fruits which animals, including humans, collect and carry away.

Because of their hard coat and dry, dormant state, seeds can survive for long periods in the soil. Common weeds such as field poppies produce seeds which may live for 30 years or more until chance brings them to the surface and they grow. The record for seed longevity is held by the Indian lotus, *Nelumbo nucifera*, seeds of which have germinated after 1000 years of lying on a lake bed.

4.40 Small seeds, such as this one from the snapdragon, *Antirrhinum majus*, are often very rough in surface texture. The ridges and craters are produced by outgrowths of the integument of the ovule. Their purpose is to trap small particles of soil, so that when the seed falls to the ground it is anchored and can germinate in security without being blown away.
SEM, ×150

4.41 Goosegrass, *Galium aparine*, disperses its seeds by means of tiny hooks which cover the outside of the fruit. The hooks here are in the process of formation. The two young fruits are still firmly attached to their common stalk. At maturity, this attachment will become tenuous and any passing furry animal, or clothed human, may be enlisted to disperse the fruit. This technique of seed dispersal inspired the invention of Velcro (see page 179).
SEM, ×140

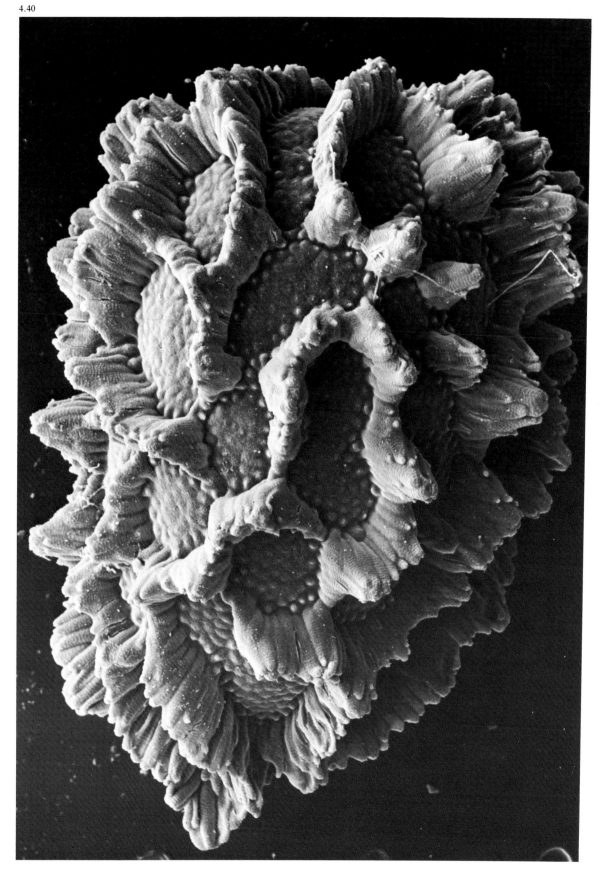

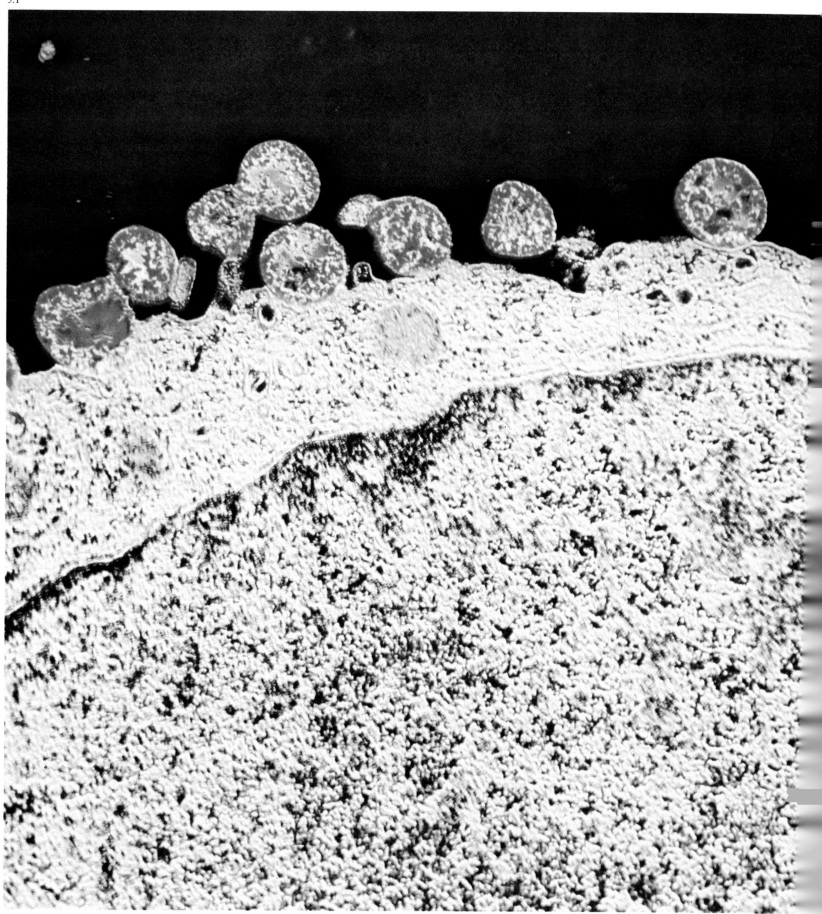

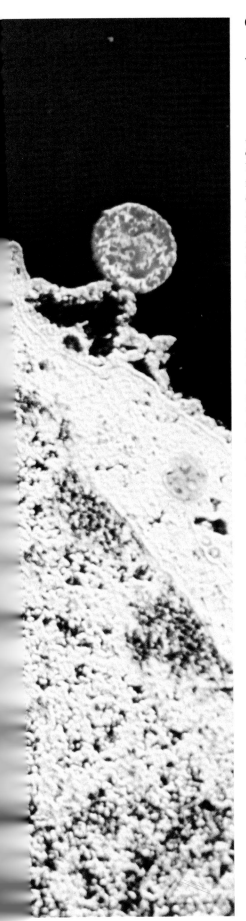

CHAPTER 5
MICRO-ORGANISMS

MICRO-ORGANISM is a popular term that is commonly applied to a wide biological spectrum. In fact, some micro-organisms are not alive, and others produce forms which are not microscopic. In this chapter, the term is used to include viruses, mycoplasmas, bacteria, algae and fungi. Protozoa have been covered already in the chapter on animals.

Micro-organisms were first seen by Antoni van Leeuwenhoek, the 17th century Dutch microscopist. He described various bacteria and protozoa, grouping them all together as 'animalcules'. Later developments have shown us that such a simple classification is not tenable. We can now see not just the external form, but details of the internal structure of micro-organisms, and their great variety is apparent. Some are indeed like 'little animals' in structure and behaviour, but some are plant-like, and in many, such as bacteria, the concept of plant or animal is irrelevant. The viruses are not living organisms at all.

Without microscopy the identification, description and study of micro-organisms would be impossible. The nature of viruses, for instance, only began to emerge after electron microscopy was first used to see them, in the late 1930s. Given a sample of micro-organisms, whether from water, soil, or diseased tissue, a biologist's first action is to look at it using a microscope.

Most micro-organisms are very small indeed. The tiniest are filamentous viruses, discovered in 1963. Consisting of a DNA molecule associated with protein, they are 5.5 nanometres $(5.5 \times 10^{-9}$ metres) thick, and about 1000 nanometres long. Other viruses form particles 20–100 nanometres across. Viruses are too small to be resolved in a light microscope. The smallest living cells, on the other hand, are just visible by optical microscopy, but need the electron microscope's high resolution to be seen in detail. Mycoplasmas and the smallest bacteria are about 0.3 micrometres $(0.3 \times 10^{-6}$ metres) across. Larger cellular micro-organisms such as yeasts and single-celled algae have dimensions of 1–100 micrometres. Some algae and fungi produce long filamentous cells which are just visible to the naked eye, and these may form large complexes. Giant kelps, for example, have fronds up to 150 metres long and are amongst the largest living things.

Scientific classification is based on biology and structure, not size. The most elementary micro-organisms are viruses, which rely on living cells for their reproduction. The smallest cells are also mostly parasitic, although some mycoplasmas and small bacteria can be cultured *in vitro* in complex nutrient media. Larger bacteria are usually free-living and adaptable; different species can survive in habitats as diverse as spring water or mineral oil.

Such simple cells are prokaryotic: their genetic information – the DNA – exists as a naked molecule within the cell. All other cells, including those of algae and fungi, are eukaryotic: their DNA is enclosed within a body called a nucleus. This distinction has evolutionary significance. The first cells to exist were almost certainly prokaryotic. Current opinion suggests that eukaryotic organisation arose by specialisation of prokaryotes acting as parasites within cells. For example, mitochondria, the small respiratory bodies found in all eukaryotic cells, may have developed from intracellular bacteria.

Micro-organisms are ubiquitous and very numerous. A gram of soil can contain 100 million bacteria and 250 000 fungal cells. Micro-organisms are in the air, in the water we drink, and living within our bodies. A few cause disease in plants or animals, but most are beneficial, particularly as scavengers of organic waste.

5.1 Mycoplasmas are the simplest living cells known; about 60 species are recognised at present. Their DNA codes for about 750 proteins, which is considered the minimum for an independent existence. They differ from bacteria in that they do not possess a cell wall. In this transmission electron micrograph, the mycoplasmas appear as red particles at the surface of an animal cell, coloured yellow. Only a fraction of the animal cell is visible; its diameter is 100 times greater than that of each mycoplasma. The brown line curving across the picture represents the boundary of the animal cell's nucleus. Mycoplasmas can cause pneumonia-like disease in humans and livestock.
TEM, stained section, false colour, ×59 400

VIRUSES

Viruses have been called mobile genes. They exist outside living cells as separate particles called virions. Each virion consists of a length of nucleic acid, the genome, together with a protein coat, the capsid. Some virions are further enveloped in a layer of lipid and protein material. Viruses are classified by shape, size and the nature of their nucleic acid.

The shape of a virus particle is determined by the arrangement of the protein molecules which comprise the coat. These subunits of the coat, called capsomeres, associate together in the form of helices or as the plane faces of polyhedra. Helical viruses may be straight rods, flexuous rods, or bullet-shaped. The commonest polyhedral form is the icosahedron, with 20 faces. The size of a virion depends on the number of capsomeres in its coat, and this in turn is related to the size of the viral genome.

The nucleic acid may be either DNA or RNA. No known virus contains both. The function of the genome is to act as a coding message for the production of enzymes which synthesise the coat protein, and copies of the genome itself. The virus is not alive; to accomplish its own replication it has to enter a living cell. Once inside its host cell, the virion's coat disintegrates to expose the genome, which proceeds to 'hijack' the cell's chemical apparatus to manufacture more coat protein and copies of itself. The new protein and nucleic acid then assemble themselves into the next generation of virions.

The transmission of virus disease can occur through the air, as in influenza, or it may need direct contact, as with rabies. Plant virus diseases are most commonly transmitted by insects with sucking mouthparts.

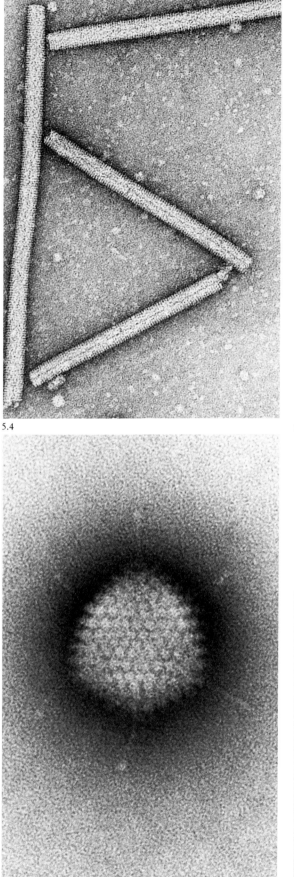

5.2

5.4

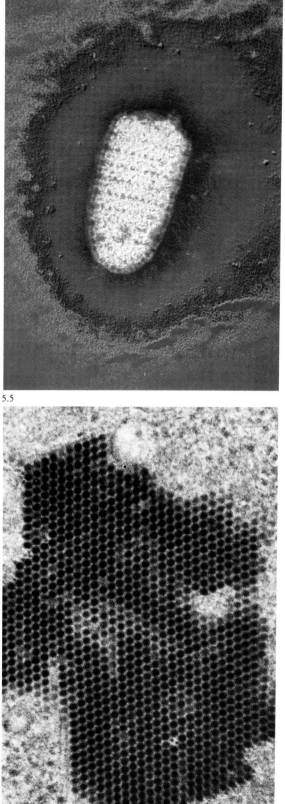

5.3

5.5

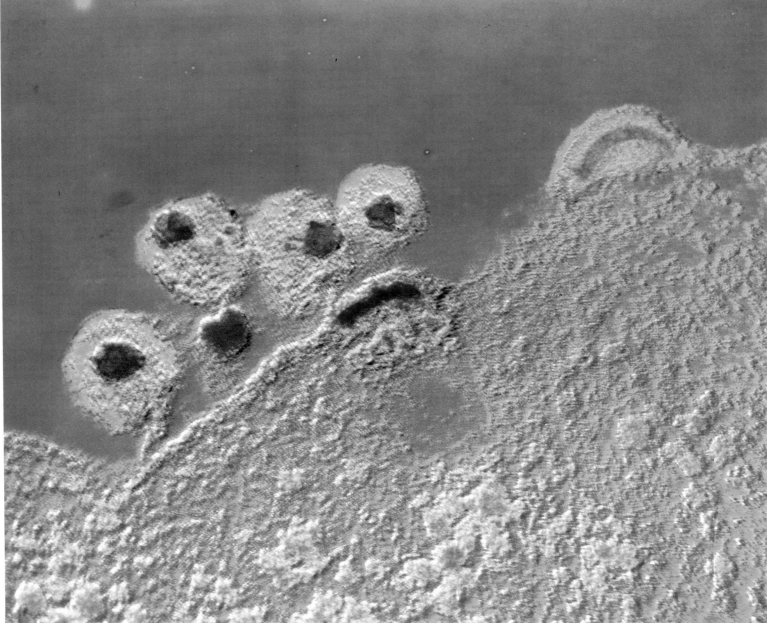

5.2 Beet Necrotic Yellow Vein (BNYV) virus is an example of a rod-shaped RNA plant virus. It is named after the symptoms which it produces on sugar-beet plants – yellowing of the leaf veins, and death. This transmission electron micrograph shows four particles of the virus, each comprising a hollow tube of capsomeres arranged in helical form. The RNA lies inside this tube. The striations show the spacing of the capsomeres, which are 2.6 nanometres apart. This clarity of very fine detail is characteristic of the negative staining technique used to make this high-magnification picture.
TEM, negative stain, ×200 000

5.3 The bullet-shaped rabies virus also has helical symmetry and a genome of RNA. The spiral arrangement of capsomeres can be seen in the light area of the micrograph. The diffuse dark surround is the viral envelope, consisting of lipoprotein material.
TEM, negative stain, ×117 500

5.4 Adenovirus contains DNA enclosed in a protein coat in the form of an icosahedron. The coat is built from 252 protein subunits. The fine spikes visible in this negatively stained preparation enable the virus to recognise the host cell. Adenovirus causes infections of the upper respiratory tract in humans, with symptoms like the common cold; it has also been implicated in cases of cancer.
TEM, negative stain, ×92 800

5.5 In this section of a cell infected with polio virus, the virus particles are visible in a crystalline array. An infected cell may produce hundreds or thousands of new virus particles. Their dark staining is due to the high concentration of nucleic acid within them.
TEM, stained section, ×83 200

5.6 AIDS (Acquired Immune Deficiency Syndrome) is caused by an RNA virus which uses as its host a white blood cell of the human immune system known as the T4 lymphocyte. The micrograph shows newly produced AIDS virus particles leaving an infected T4 cell. The deep red core of each particle corresponds to the viral RNA. The new particles will go on to spread the infection, which results in the destruction of the immune response. The victim eventually dies of a secondary infection which his or her immune system is unable to fight.
TEM, stained section, false colour, ×191 000

BACTERIOPHAGES

Bacteriophages are viruses which infect bacteria. They have played an important part in our understanding of genetic processes and in the development of molecular biology. Like other viruses, they consist of a piece of nucleic acid enclosed in a complex protein coat.

The replication cycle of a bacteriophage begins with its attachment to the outer surface of the host bacterium. The nucleic acid within the virus is injected into the host, often by means of a contractile tail. Only the nucleic acid enters the bacterial cell, and this was an important demonstration in the 1950s that DNA on its own is sufficient to encode the structure of complete virus particles. Once inside the cell, the DNA can do one of two things.

In a virulent infection, it immediately starts to direct the synthesis of proteins, including the enzymes concerned with the synthesis of new DNA. Thus the virus circumvents the cell's normal control processes, which ensure that DNA is replicated only once in each generation. With its own enzymes in control, the virus can replicate itself hundreds of times, and within 30 minutes or so the bacterium becomes filled with new virus particles. It then bursts, releasing the new viruses.

In a lysogenic infection, the viral DNA inserts itself into the bacterial chromosome. There it waits, being replicated in the normal way each time the bacterium divides. Thus a single bacterium can give rise to dozens or hundreds of infected progeny, each carrying the virus in latent form. Only when these bacteria are subjected to a life-threatening stress, such as ultraviolet light or ionising radiation, does the viral DNA leave the chromosome and

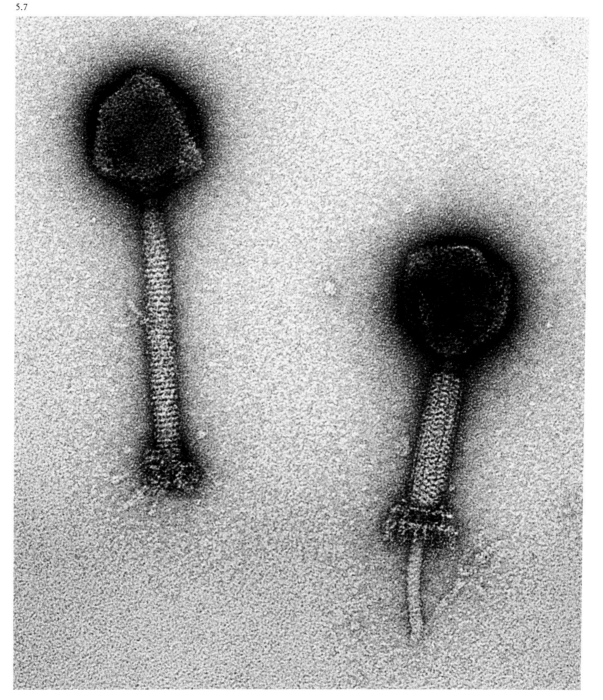

immediately direct the formation of new virus particles. In this way the virus, but not the bacterium, survives.

Bacteriophages can be used in DNA cloning. A novel piece of DNA is inserted into the viral DNA, and infection allowed to proceed. The host bacterium divides to eventually produce thousands of copies of the inserted DNA. This technique is used for identification and isolation of useful genes, such as those

concerned with the production of antibiotics and human hormones.

Bacteriophages occur naturally in a wide range of environments – in soil, sewage, dairy products and diseased tissue. Unlike the viruses of higher organisms, which are commonly named after the disease they cause, bacteriophages are identified by letters and numbers.

5.7 The complex protein capsule of these two SP105 bacteriophages is in three parts. The large head region

houses most of the DNA. The tail is contractile and serves to inject the DNA into the host through a tubular core. The fine fibres at the end of the tail recognise and bind to the surface of the host bacterium prior to infection. In this transmission electron micrograph, the bacteriophage on the left has its tail in expanded mode, while the one on the right has its tail contracted, revealing the core through which the DNA has travelled.
TEM, negative stain, ×470 000

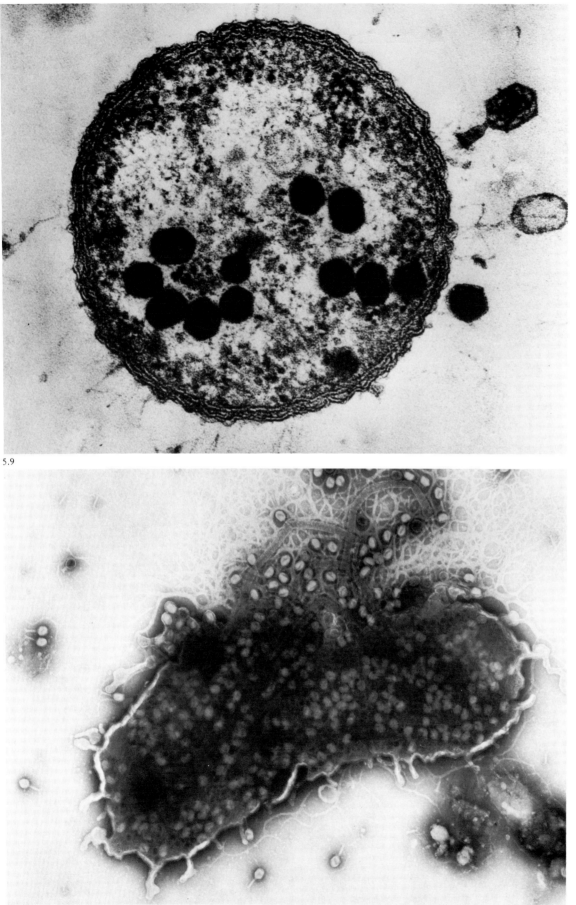

5.9

5.8 This section through a cell of the bacterium *Escherichia coli* shows it being attacked by T2 bacteriophages. Nucleic acids stain easily and appear dark in electron micrographs. The empty T2 particle at right has already injected its DNA into the bacterium, and new virus particles are visible as the solid black objects within the cell. The much smaller granular particles in the cell interior are ribosomes, on which protein synthesis takes place. A section like this is very thin – about 70 nanometres thick. This is about 1/50th of the complete bacterial cell.
TEM, stained section, magnification unknown

5.9 The bursting, or lysis, of the host bacterium is the final stage of a virulent infection. Graphically captured in this micrograph, the process results in the release of thousands of new bacteriophage particles. The ones in this picture are T4; they appear as small white particles because the specimen has been negatively stained. The specimen consists of the whole bacterium, not a thin section as in the previous picture.
TEM, negative stain, ×120 000

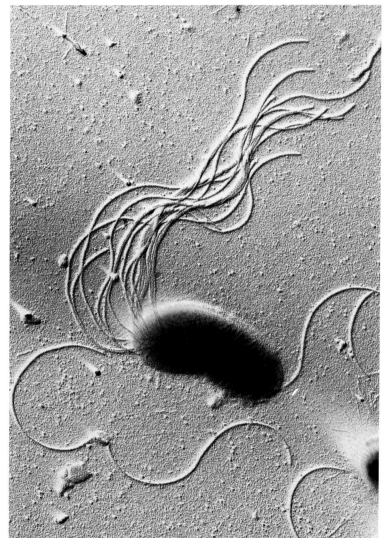

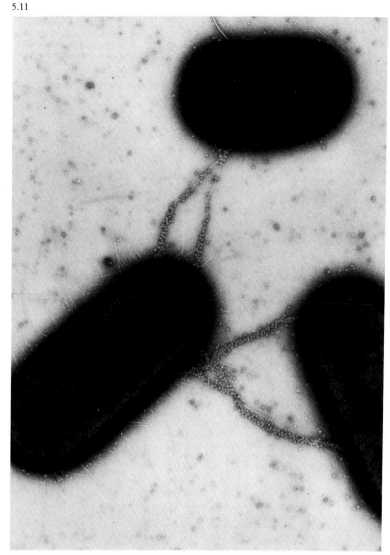

BACTERIA

Bacteria are prokaryotic cellular micro-organisms. There are about 2000 species, ranging in size from spheres 0.3 micrometres across to filamentous cells 20 micrometres long. Bacteria come in three basic shapes, as spheres (cocci), rods (bacilli), or spirals (spirilla). Individual bacteria may associate into groups; the Streptococci, for example, form chains of spherical cells.

The bacterial cell is elementary in structure. It contains DNA in the form of a closed loop as its genetic material. The cytoplasm is filled with ribosomes – the particles on which protein synthesis takes place. The cell membrane may have folded regions, known as mesosomes, on its inside surface. The external surface may be smooth, covered with a slimy capsule, or it may have fine hairs, called pili. These attach the cell to its substrate, and to other cells during sexual union. Motile bacteria have one or more flagellae. These sinuous protein filaments rotate to propel the cell through liquid environments.

Bacteria differ in their ability to use nutrients. The heterotrophs require preformed organic molecules produced by other cell types, and they are therefore associated with other organisms – occurring, for example, in the human mouth and gut. Autotrophs can use inorganic minerals and atmospheric carbon dioxide to satisfy their needs. They are widespread in the soil and water.

A few species of bacteria cause illness, such as the agents of cholera, plague and legionnaire's disease. But the vast majority are beneficial, acting to break down organic wastes, maintain soil fertility, and even digest some components of our food for us.

5.10 To obtain this picture, a cell of the soil bacterium *Pseudomonas fluorescens* was coated with a layer of carbon followed by platinum. This produces a surface replica which is used as the microscope specimen. *P. fluorescens* is motile, and uses the bunch of whip-like flagellae seen here to move through the water layer surrounding soil particles.
TEM, surface replica, ×18 750

5.11 Sex between bacteria, known as conjugation, was discovered in 1946. It involves the transfer of DNA between cells. In this picture, three cells of *Escherichia coli* are visible. The one at bottom left is male, the other two are females. Maleness in *E. coli* is associated with the presence of long, hollow surface hairs, called F-pili. The male has attached F-pili to each of the female *E. coli*, and DNA is transferred between them through the hairs' hollow cores. Careful control and interruption of this process enables scientists to map the relative positions of the bacterium's genes. The tiny granular particles that almost completely cover the F-pili in this micrograph are particles of MS2 bacteriophage, which bind specifically to the F-pili.
TEM, negative stain, ×10 250

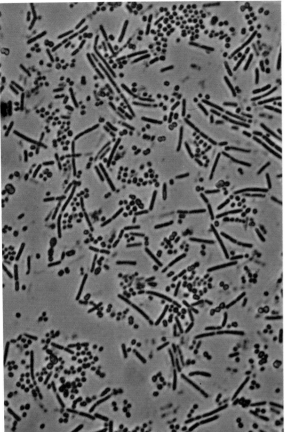

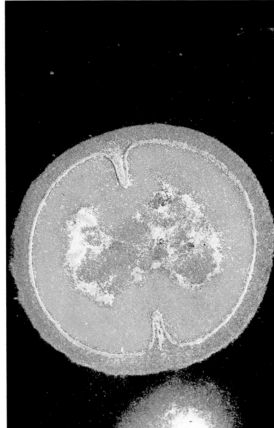

5.12 This light micrograph of a population of bacteria from the human mouth demonstrates the limitation of optical microscopy in the study of such small cells. Compared to the other pictures on these pages, there is very little detail visible beyond the outline shape of the bacteria, a mixture of cocci and bacilli.
LM, bright field illumination, ×1600

5.13 Bacteria divide by splitting in two, a process called binary fission. In this picture of a dividing cell of *Staphylococcus epidermidis*, the blue and yellow region in the cell centre is the genetic material, about to break into two parts. The division is completed by the ingrowth of a new cell wall, seen here as the fine yellow line around the edge of the cell. The blue layer beyond is the external wall covering. *S. epidermidis* is found in quantity all over human skin, where it is generally non-pathogenic, though it may play some part in causing acne and in making cuts and scratches go septic.
TEM, stained section, false colour, ×60 000

5.14 This bacterium, a species of *Leptospira*, takes the form of a long thin spiral. *Leptospira* is one of a group of helical bacteria known as spirochaetes, which contains many dangerous pathogens. The condition leptospirosis is commonly known as Weill's disease, and is transmitted by contact with rats. This single cell, in red, is about 20 times larger than the other bacteria on these two pages.
TEM, negative stain, false colour, ×20 000

5.14

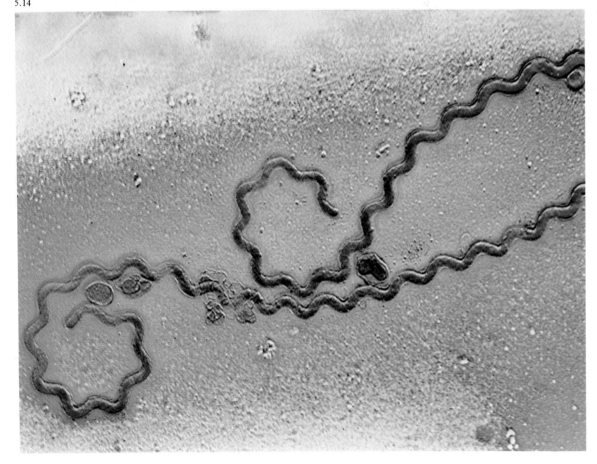

E. COLI

Escherichia coli, or *E. coli* as it is commonly known, is a normal and usually harmless inhabitant of the intestinal tract. It is also the world's premier laboratory organism. A vast amount is known of its genetics, and many laboratories house large collections of different strains, each capable of carrying out a specific series of reactions. Much of our understanding of how genes work comes from the study of this bacterium.

Today this detailed knowledge is exploited by the techniques of genetic engineering. By deliberately introducing pieces of foreign DNA into the bacterium, it is possible to make it produce useful medical products such as insulin or interferon.

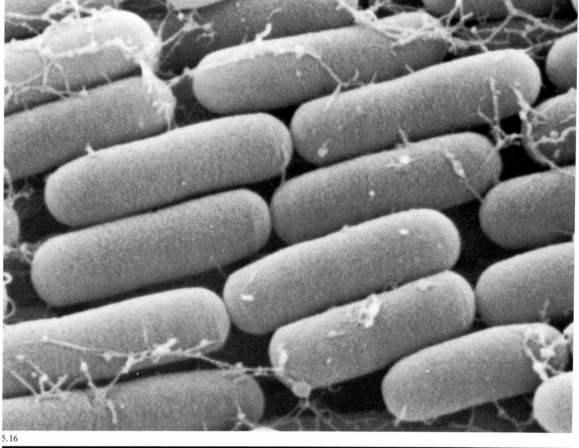

5.15 In the scanning electron microscope, the image is strongly three-dimensional in appearance, and the *E. coli* cells are seen as typical rod-shaped bacilli. The fibrous material is the remnants of the nutrient medium in which the bacteria were grown. SEM, ×17 000

5.16 By contrast, the transmission electron microscope gives an image of the internal details of cells, with little three-dimensional content. This picture is of a section of *E. coli* only 70 nanometres thick. The black regions are DNA which has duplicated itself in preparation for binary fission. TEM, stained section, ×78 000

5.17 To make this picture, the bacterial cell in the centre was treated with an enzyme to weaken its wall, then placed in water, causing its DNA to be ejected. It is visible as the gold-coloured fibrous mass lying around the cell, and its length is 1.5 millimetres, or 1000 times the length of the cell from which it came. The specimen has been shadowed with a layer of platinum, a technique which gives very high resolution pictures. TEM, shadowed replica, false colour, ×67 500

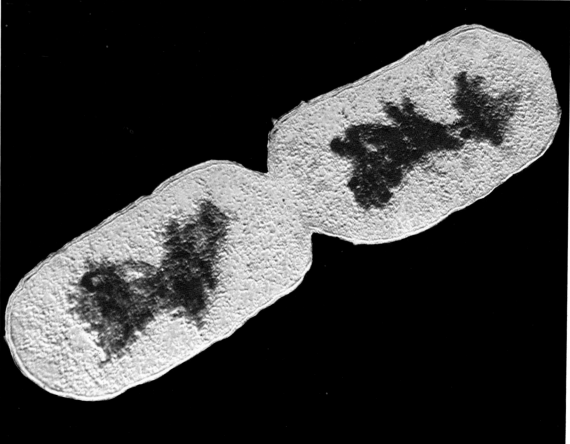

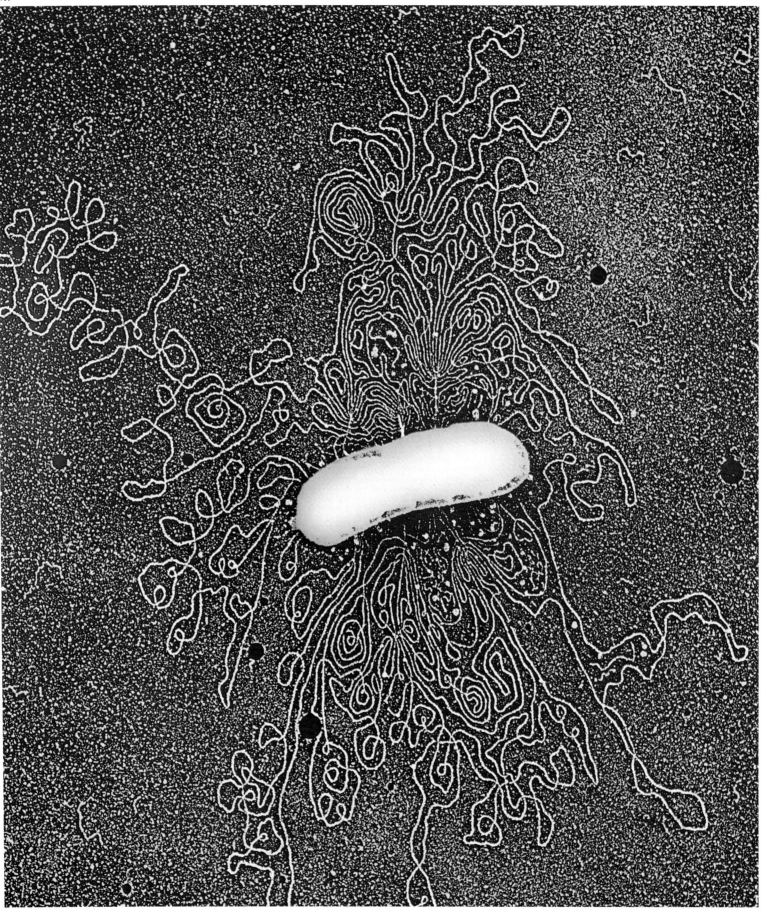

RHIZOBIUM

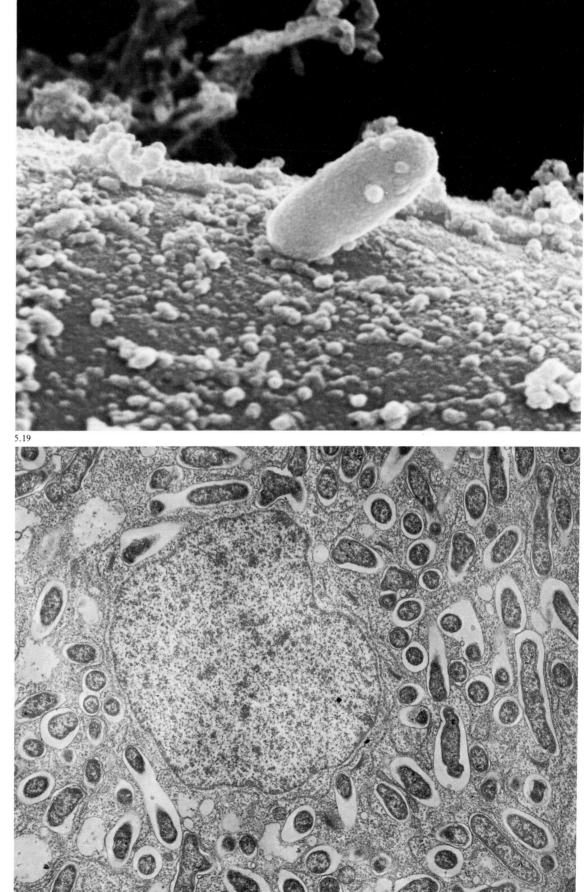

Fertile soils contain many species of bacteria, most of which break down organic materials and create humus. One which has a more directly beneficial effect on plant growth is called *Rhizobium leguminosarum*. It forms a specific relationship with the roots of leguminous plants, a family which includes clover, peas and beans.

The first contact in the relationship is between a bacterial cell living free in the soil and one of the fine hairs which the plant's roots produce in order to extract moisture from the ground. In response to the contact, the root hair admits the bacterium to the interior of the root by means of a specially constructed tunnel called an infection thread. Once inside the root, the bacterium causes the root cells to divide. This eventually produces a large outgrowth from the root surface, called a nodule, which is visible to the naked eye. Within the cells of the nodule, the bacterium loses its outer wall, and divides repeatedly to produce a mass of new cells which, because they lack cell walls, are called bacteroids. Thus the plant acts as an amenable host for the multiplication of a single bacterium into millions of bacteroids.

The relationship, however, is not one of parasitism, but of symbiosis – a partnership beneficial to both sides. In return for its place of residence in the nodule, *Rhizobium* performs a unique service for the plant. In its bacteroid form, it can transform the nitrogen gas in the air which permeates the soil into ammonium salts. These act as nitrogenous fertiliser for the plant.

Rhizobium is by far the most important of the soil bacteria which fix nitrogen for plant use. It has been estimated that 150–200 million tonnes of nitrogen are fixed

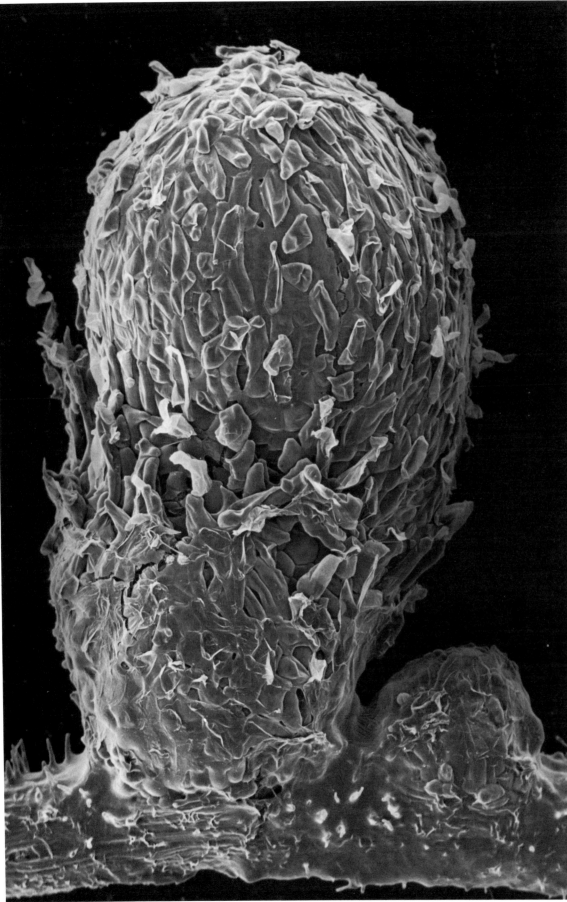

by bacteria annually – about three times the total world production of nitrogen fertilisers from chemical factories. The life cycle of *Rhizobium* is completed when, after a period of a few weeks, the nodule breaks down. The bacteroids then form cell walls and enter the soil as bacteria, to repeat the process of infection.

5.18 This high-magnification scanning electron micrograph shows a single bacterium attached to a root hair of a laboratory-grown garden pea plant. The attachment is very specific: *Rhizobium leguminosarum* infects only leguminous plants, although it is not known how it recognises them; and the initial contact with the root hair often appears to occur at one end of the rod-shaped bacterium, as in this picture. The granular particles on the surface of the root hair and the bacterium are remnants of the medium in which the plant was grown.
SEM, ×40 000

5.19 Each infected cell is host to thousands of bacteroids. In this section, the large central body is the nucleus of a nodule cell. Around it are ranged the small, dark-staining bacteroids; of indefinite shape, they are separated from the host cell by a membrane and a clear space. Half the weight of a mature nodule may consist of bacteroids.
TEM, stained section, ×6250

5.20 This large nodule is about 3 millimetres in diameter. The loosely packed cells on its surface are not normally as infected with bacteroids as internal cells like the one in Figure 5.19. The nodule has grown from the length of pea root at the bottom of the picture. To the left of the nodule's base can be seen a few root hairs; to its right is a much smaller nodule.
SEM, ×35

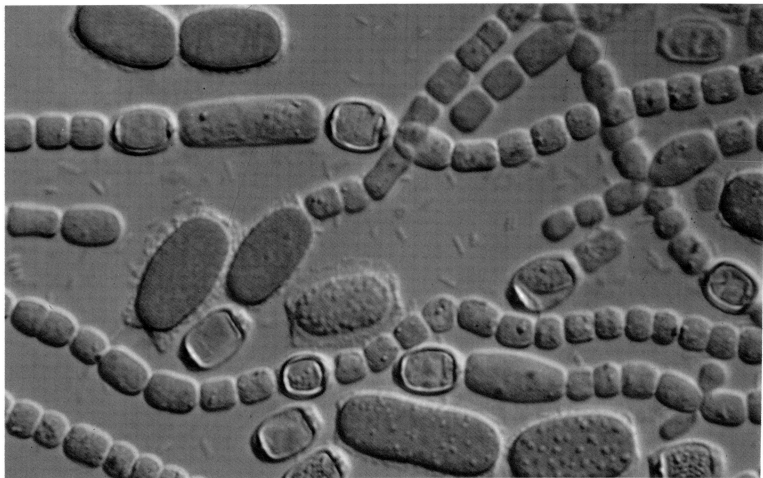

ALGAE

Algae are found wherever there is water and light. They all share the ability to transform carbon dioxide into sugars by the process of photosynthesis. Algae have been called 'the grass of the waters'; they are immensely important as the base of many food chains, particularly in the oceans. Algal marine plankton fixes 10^{10} tonnes of carbon each year by photosynthesis – more than the total production of all the world's land plants.

Most algae are microscopic, existing as single cells or filaments of cells joined end to end. Some, such as seaweeds, form large visible bodies. The life cycles of algae are often complex, with highly specialised sexual reproduction, which takes place in water.

Algae use mixtures of pigments to perform photosynthesis, and this is one basis of a broad classification into classes – the green algae, brown algae, red algae, and so on. The cells of these algae are eukaryotic, with a distinct nucleus and the pigments localised in organelles called chloroplasts. Blue-green algae are primitive relatives, sometimes classified with bacteria. Like bacteria, they are prokaryotic, but their main pigment is chlorophyll, which is found in algae and all higher land plants. It is possible that the chloroplasts of higher plant forms arose originally from blue-green algal cells which were ingested by a non-photosynthetic cell early in evolution

5.21 Different cell types may form even in primitive organisms such as this blue-green alga, *Cylindrospermum*. The chains of small units in the light micrograph are vegetative cells which perform photosynthesis. The slightly larger cells with thick walls are called heterocysts, and they can fix nitrogen. The largest cells are a type of spore called an akinete.
LM, Nomarski DIC, magnification unknown

5.22 The desmids are green algae characterised by their beautifully shaped cell walls. This is a single cell of the desmid *Micrasterias*, seen among a variety of filamentous algae and debris. The cell is in two halves, separated by a narrow waist. Cell division is accomplished by splitting in two at the waist, with each half generating a perfect replica of itself to restore the original shape. The cell in this picture divided some time ago, and the new half cell (upper right) is almost full-grown.
LM, spectral Rheinberg illumination, ×850

5.23 The green alga *Volvox* forms spherical colonies consisting of hundreds or thousands of cells glued together to form a hollow sphere. The individual cells appear as bright green particles in the picture. The larger green masses, of which six are visible, are asexual daughter colonies forming inside the sphere.
LM, dark field illumination, ×315

5.24 *Spirogyra* is a green alga in which the cells join end to end to form filaments. Its chloroplasts are arranged in spiral patterns which are disrupted during cell division. In this picture the cell in the middle of the frame is dividing; those above and below it show the spiral chloroplasts typical of non-dividing cells.
LM, bright field illumination, ×200

5.25 *Hydrodictyon* is another colonial green alga. Colonies can comprise up to 20 000 cells joined so as to make a hollow floating net which is closed at both ends. Large colonies may reach 50 centimetres in length. This picture of a small part of a colony shows how the individual cells join and branch at their ends to produce the three-dimensional network.
LM, Rheinberg illumination, ×735

5.22

5.23

5.24

5.25

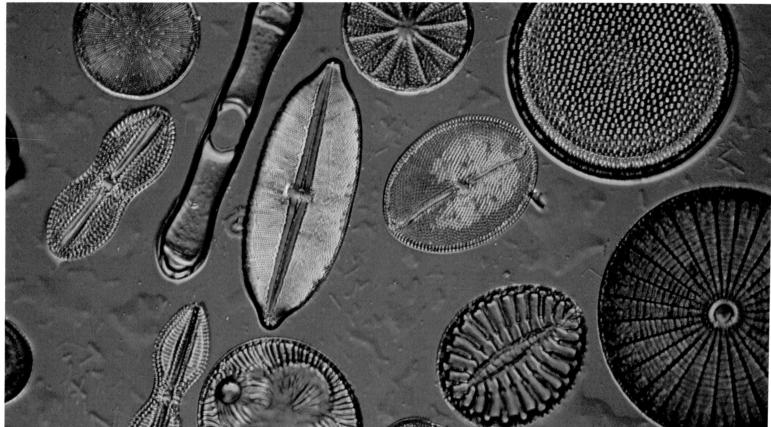

DIATOMS

The diatoms are a distinctive group of single-celled algae. Comprising about 10 000 species, they form an important part of the plankton in fresh and salt waters. The number of diatoms in the seas is immense, especially in temperate latitudes, where a litre of surface sea water may contain as many as 15 000 diatoms.

The characteristic feature of diatoms is their intricately patterned, glass-like cell wall, or frustule. The frustule consists of two halves, the valves, which fit together like the two parts of a pill box. One half is slightly larger than the other, and acts as the 'lid' of the box.

When a diatom divides, each valve produces a new half inside itself. In consequence of this, the 'lid' generates a cell which is identical in size to the original one, whereas the 'bottom' gives rise to a cell which is slightly smaller. This process can be repeated many times, but eventually the ever smaller progeny of the 'bottom' will cease to be viable. The diatom solves this strange paradox by occasionally producing sexual 'auxospores' which restore the maximum cell size.

Diatoms have long been favourite subjects for microscopists. Victorian amateurs amused themselves by arranging diatoms in complex patterns. The frustule, made from silicates which the diatom extracts from the water around it, is often decorated with tiny holes arranged in rows (*striae*) so fine and regular that they have been used as test objects to measure the quality of microscope lenses. In industry, diatomaceous earth formed from skeletons of countless billions of diatoms is used in products as diverse as toothpaste, dynamite and sealing wax.

5.26 Diatoms are classified into two broad groups depending on their structure. In centric diatoms, the striae are arranged radially. In pennate diatoms, the striae are in rows on either side of a central axis of symmetry. The picture shows representatives of both types. The colours are due to the use of polarised light, not to any pigmentation of the frustule.
LM, Nomarski DIC, ×280

5.27 Some diatoms form very simple colonies consisting of cells which have failed to separate following division. This picture shows 14 such cells of the species *Fragillaria crotonensis*. Each one is glued by its centre to its neighbours with a mucilaginous material. Because of the mechanism of division, the cells at either end of this colony are different in size, but the difference is too small to detect with certainty. *F. crotonensis* lives in freshwater lakes.
SEM, magnification unknown

5.28 Many diatoms cannot move independently; those that can usually achieve only a strange jerky motion. *Navicula monilifera* is one such motile species. A pennate diatom, its striae are arranged in rows on either side of a central furrow, the raphe. The raphe is the secret of the diatom's mobility. Fluid is squirted along it and pushes the whole cell forwards. The power of movement is most common in bottom-living species such as *N. monilifera*; diatoms which float rely on water currents to carry them from place to place.
SEM, ×1250

5.29 The decorated valves of diatoms consist of a mixture of pectin and silicates. In centric diatoms, such as *Cyclotella meneghiniana*, the two valves are separated by an undecorated girdle band. This can expand, allowing the cell to grow. *C. meneghiniana* inhabits the brackish waters of salt lakes and estuaries.
SEM, ×3750

5.30 This cell of *Biddulphia* is lying on its side, and gives a clear view of the girdle band separating the two valves. The band holds the valves together while at the same time permitting cell growth. *Biddulphia* is a component of marine plankton.
SEM, ×1200

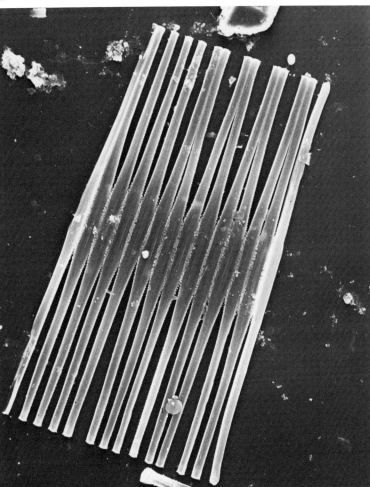

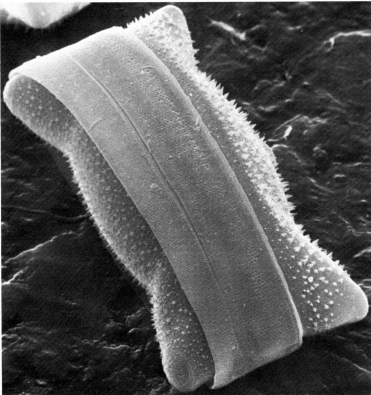

FUNGI

Fungi, like algae, occur in a variety of forms ranging from truly microscopic organisms up to visible and familiar ones such as mushrooms and toadstools. Many exist in the form of single cells. The higher fungi characteristically produce a filamentous growth called a hypha, which branches repeatedly to form a network known as a mycelium. Fungi do not form true tissues. The 'stem' of a mushroom, for example, consists of a tangled mass of hyphae.

The fundamental difference between fungi and algae is the complete absence of chlorophyll in fungi. As a result, fungi cannot perform photosynthesis, and have to rely for sustenance on organic material produced by some other living creature.

Many fungi feed on the dead remains of other life forms. Known as saprophytes, their favourite haunt is the forest floor with its supply of fallen leaves and branches. Mushrooms and toadstools are saprophytes. Saprophytic fungi perform a useful function in the maintenance of soil fertility by contributing to the formation of humus. They are also familiar as the moulds which spoil our food; they are quite at home in a pot of jam, or on the surface of bread, cheese, or an orange. Other representatives live off the fabric of our houses – dry rot and wet rot of timber are both caused by saprophytic fungi.

The other major fungal lifestyle is parasitic: the fungus attacks the living tissue of its host, causing symptoms of disease. It was the impact of a fungus disease – potato blight in 19th century Ireland – which first gave impetus to the science of plant pathology. Fungal diseases of humans are by and large less severe in their effects than bacterial or viral infections.

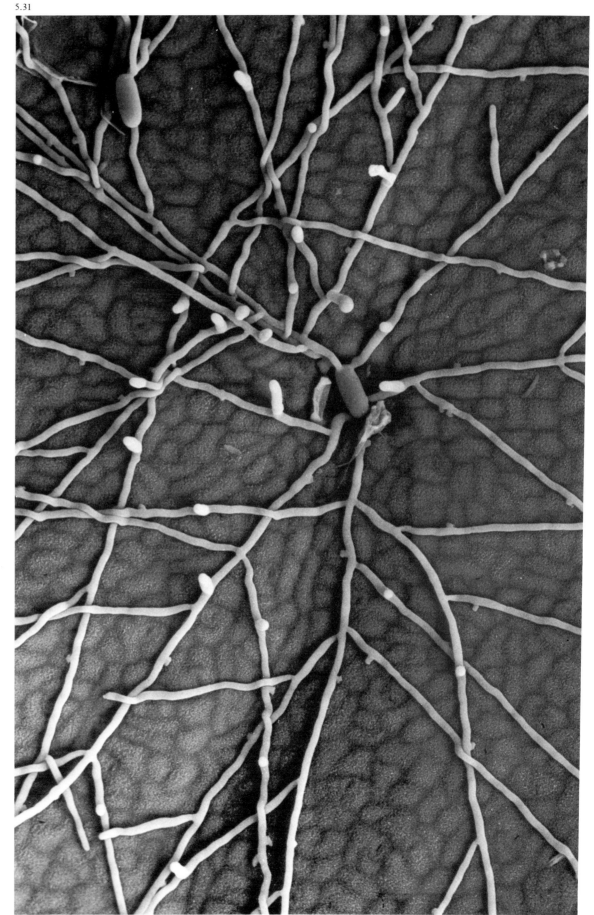

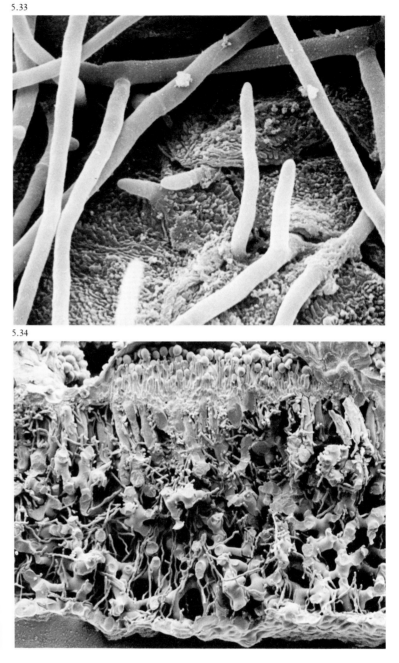

5.34

5.31 One of the commonest plant diseases caused by fungi is powdery mildew. This generic name refers to the appearance of an infected plant, the leaves of which seem to have been dusted with powder. The powder in fact consists of the tiny spores of the fungus. A wide range of fungal species are referred to as 'powdery mildews'. The one depicted here is *Erisyphe pisi*, which infects garden peas. The infection begins with a spore landing on a leaf surface. The originating spore can be seen just above centre in this picture. The spore germinates and produces a series of branching filaments, which creep over the leaf surface and at intervals send branches down into the leaf interior. Eventually the fungus develops aerial branches which give rise to more spores. These are blown away by the slightest breeze to spread the infection. Crops such as peas and cereals, and many grasses, are affected by powdery mildews.
SEM, ×280

5.32 The spore-bearing structures (sporangia) of fungi are often elaborate in shape, and can be used as a means of identification. This one, from *Mycotypha africana*, is shaped like a bottle brush. The picture also shows the hyphal filaments of this fungus, which is another powdery mildew.
SEM, ×900

5.33 Skin and hair are rich sources of a protein called keratin. This is used as a food source by a number of fungal species which cause the symptoms of ringworm and athlete's foot. Here, the hyphae of *Trichophyton interdigitalis* are shown growing through the epidermal scales of human skin. The fungus lives in the soil and on small furry animals such as voles; in humans it causes ringworm.
SEM, ×3500

5.34 To see what is happening inside a leaf, it is necessary to break it open. This scanning electron micrograph was obtained by freezing a leaf from a bean plant infected with a rust fungus, *Uromyces fabae*, and then snapping it in half. The view is of the broken edge, and it shows that the interior of the leaf is infested with string-like hyphae of the fungus. Rusts are so called because they produce brown pustules on the surface of infected leaves. These rust spots contain millions of spores ready to spread the infection. A mass of such spores is visible on the surface of the leaf at the top of the picture.
SEM, ×145

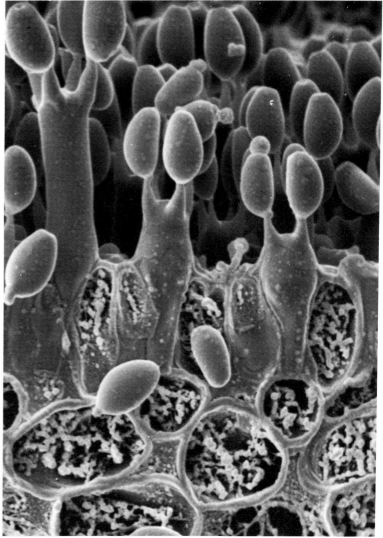

5.35 *Penicillium* species are the most widespread moulds known to humanity: the air is full of their spores, called conidia. As a result, exposed items of suitable foodstuffs such as bread, milk and cheese quickly become infected with the fungus. Initially, growth is in the form of colourless hyphae, but after a few days the fungus produces special aerial branches called conidiophores, seen here growing on a mouldy piece of Cheddar cheese. At the ends of the conidiophores rows of conidia are formed. They are green in colour, and this is why mouldy cheese often looks green. The slightest movement of air causes the conidia to be detached and blown away. It was the chance arrival of a spore of *Penicillium notatum* on a bacterial plate in Alexander Fleming's laboratory that led to the discovery and eventual purification of the antibiotic penicillin. In nature, penicillin is used by the fungus to suppress the growth of bacteria which might compete with it for food, or use it as food. Commercial antibiotic production uses *P. chrysogenum*. Other *Penicillium* species are used in the manufacture of cheeses such as Roquefort and Camembert.
SEM, ×400

5.36 The familiar mushroom or toadstool is only a small part of the life cycle of the fungus which produces it. It results from years of growth of the mycelium in the soil. When conditions are right, usually in the autumn in temperate regions, parts of the mycelium associate together to form a compacted tissue mass which lifts itself out of the ground. This 'fruit body' – the visible mushroom or toadstool – produces millions of special cells called basidia from a surface layer known as the hymenium. In this scanning electron micrograph of the fruit body of *Coprinus disseminatus*, the basidia are the elliptical cells at the top of the picture. The bottom of the picture shows a cross-section of the hymenial layer. Sexual union takes place inside the basidia, and this results in the formation of small cells called basidiospores, which are released and carried away by air currents to found new colonies of the fungus. After release of the spores, the fruit body breaks down and disappears, but the mycelium remains alive and will produce more fruit bodies in subsequent years. *C. disseminatus* is known as the 'crumble cap'. It is edible, but very small, and grows on rotting tree stumps.
SEM, ×4000

5.37 Brewer's yeast, *Saccharomyces cerevisiae*, is a single-celled fungus which divides by budding. Several small buds are visible on the yeast cells in this scanning electron micrograph. Yeast has been used in the production of alcoholic drinks for nearly 5000 years. The alcohol is produced as a result of the yeast feeding on sugar in the absence of air. In beer-making, the sugar comes from the germinating grain of barley. Wines are produced using a different species of yeast, *S. ellipsoideus*, which grows naturally on the skin of grapes. Yeast is also used in bread-making. Within kneaded dough there is plenty of air, so the yeast does not produce alcohol. Its respiration results, however, in the formation of carbon dioxide gas, and it is this which causes the dough to rise. Yeasts are grown commercially and harvested for the production of vitamin B1, riboflavin and nicotinic acid. Under favourable conditions of culture, a yeast cell can turn into two in the space of about 100 minutes.
SEM, ×2550

CHAPTER 6
THE CELL

ALL living things are made from fundamental units called cells. A cell is a compartment bounded by a membrane – the plasma membrane – which controls the flow of materials between the cell and its environment. Many organisms consist of single cells – for example, the protozoa. Larger creatures contain thousands, millions, or billions of cells. Just one drop of blood from a pricked finger contains about five million cells.

The word cell was coined in its scientific use by Robert Hooke, to describe the compartments in a slice of cork. Although light microscopy is adequate for studying whole cells, it was not until the invention of the electron microscope – and techniques for cutting very thin sections – that the complexity of their internal structure was appreciated. Cells are classified into two types, prokaryotic and eukaryotic, depending upon their internal organisation. This chapter is concerned with eukaryotic cells: those whose genetic material is concentrated in a distinct *nucleus*.

The interior of the cell, bounded by the plasma membrane, is filled with a fluid called cytoplasm. Within the cytoplasm are various bodies, known as cell organelles, which are also enclosed by membranes. Principal of these is the nucleus, with its content of genetic material. All eukaryotic cells also contain mitochondria, which are organelles concerned with energy production and respiration. Plant (but not animal) cells contain a family of organelles known as plastids, of which one member, the chloroplast, is essential for photosynthesis. All cells also contain a variety of other membranes in the cytoplasm,

collectively called the endomembrane system. These are concerned with the chemical synthesis of polymers such as proteins, and their transport within the cell.

The fluid of the cytoplasm was formerly regarded as a more or less uniform mixture of soluble chemicals peppered with tiny particles, the ribosomes. It is now known to be permeated with a network of protein filaments, which are thought to act as a 'skeleton' for the ordered movement of chemicals and organelles.

During growth, the number of cells in an organism increases by cell division. After division, cells usually adopt a specialised function. In the human body, about 200 different cell types can be distinguished on the basis of their structure and biochemical speciality. In plants the number of different cell types is about 20.

6.1

6.1 One of the characteristics that distinguishes plant cells from animal cells is that plant cells have an additional external envelope outside the plasma membrane. Made from a mixture of proteins and polysaccharides, including cellulose, it is known as the plant cell wall. In this transmission electron micrograph of a group of cells inside the root tip of a maize plant, *Zea mays*, the wall appears as a thin layer between the cells. The wall defines the shape of the cell, and it expands as the cell grows. The prominent round organelle in each cell is the nucleus; and each nucleus, in turn, contains a smaller, dark-staining body called the nucleolus. The white areas in the cytoplasm are vacuoles – water-filled spaces which expand and coalesce during cell growth. The grey bodies in the cytoplasm are mitochondria (pale grey) and plastids (darker grey). The small white ovals within the plastids are starch. These particular root cells form part of the meristem – the region which functions to produce new cells continually as the root elongates. Their

'speciality' is to divide; similar cells are found in shoot tips and under the bark of woody stems, such as the trunk and branches of trees.
TEM, stained section ×5000

6.2 Unlike plant cells, animal cells do not have a rigid wall; their plasma membrane does not impose a fixed shape on the cell. In this section through a human lymphocyte, the cytoplasm is coloured green. The large nucleus (orange-brown) occupies most of the internal space of the cell. The orange spot near the bottom of the nucleus is the nucleolus, while the darker brown material around the inside edge of the nuclear membrane is the genetic material, called chromatin. The smaller organelles in the cytoplasm, also orange-brown in colour, are mitochondria. Lymphocytes are part of the immune system, and their function is to produce antibodies. This involves the activation of the cell, which expands from its resting state shown here.
TEM, stained section, false colour, ×26 000

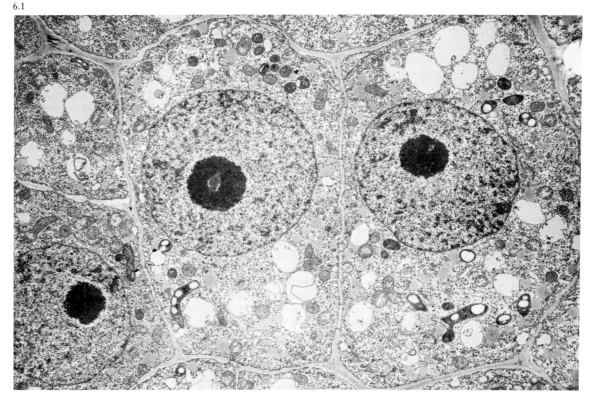

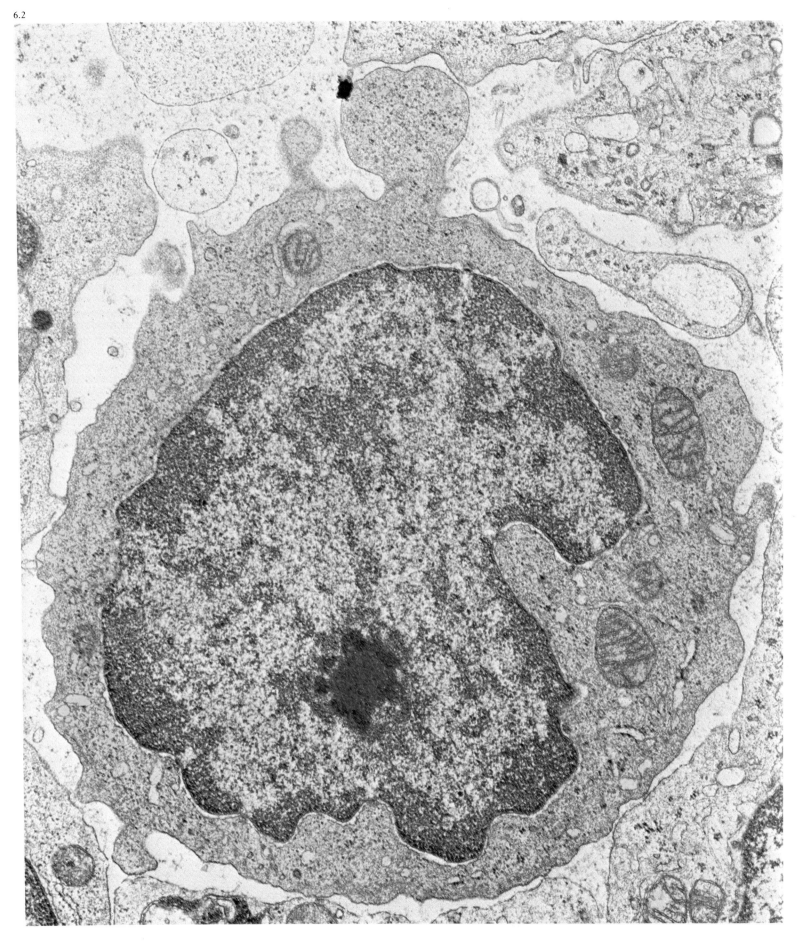

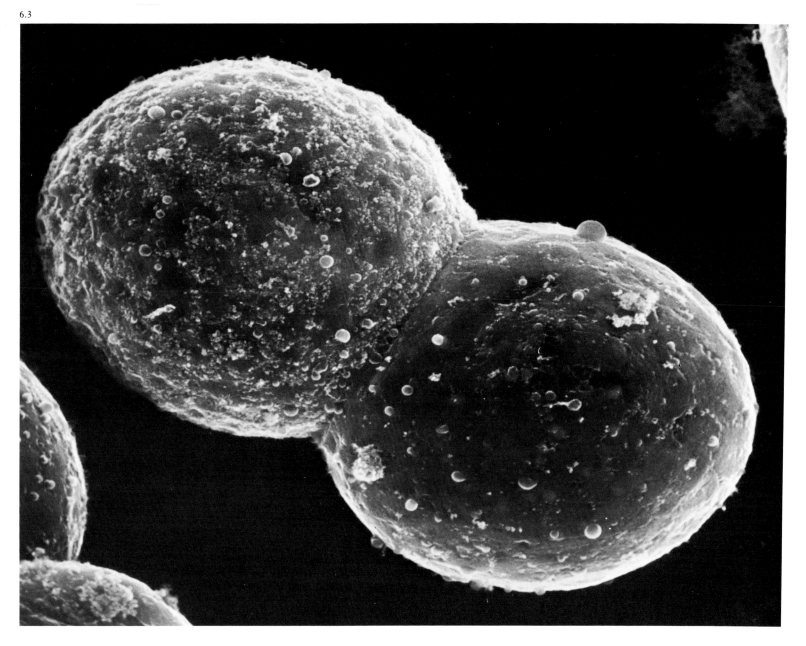

6.3 A plant cell within its wall is like a balloon inside a box: its shape is determined by the box, and it is protected from external damage. If the cell wall is dissolved away, the cell becomes spherical. In this artificial state it is called a 'protoplast'. Protoplasts are very delicate and have to be kept in a special nutrient medium to prevent their bursting, but they have several properties of great interest to biologists. One is their ability to reform their cell walls and, in certain cases, to grow back into normal plants. Another is the fact that, because their plasma membrane is exposed, protoplasts can be fused together by means of chemicals or electrical impulses. This scanning electron micrograph shows two protoplasts, derived from leaf cells of a tobacco plant, *Nicotiana tabacum*, which are in the act of fusing together as a result of treatment with the chemical, polyethylene glycol. The final fusion product will be a single cell which contains two sets of genes, one from each protoplast. This is the reason for the interest in protoplast fusion: if two protoplasts from different plants can be fused together, and a new plant grown from this fusion product, the result will be a hybrid. And because protoplast fusion can be artificially triggered, it is not subject to the natural barriers which limit other methods of plant breeding. It is theoretically possible, therefore, to create entirely new hybrid plants. Although protoplast techniques are comparatively new – the first experiments were performed in the late 1960s – some advances have been made in this direction. Another advantage of protoplasts is that they present plant breeders and genetic engineers with a means of handling large numbers of plant cells in uniform solution. This allows experiments to be carried out – for example, on resistance to virus disease – which would be impossible using whole plant tissues.

SEM, critical point dried specimen, ×2100

NUCLEUS

The nucleus is the site of most of the cell's genetic material or DNA. Nuclear DNA is always associated with proteins to form a complex called chromatin which exists in paired pieces, each of which is known as a chromosome. The number of chromosomes is determined by species – human cells, for example, have 46 chromosomes in 23 pairs. When a cell is not dividing, the DNA is in an extended form and not usually visible as discrete chromosomes.

The DNA's function is to tell the cell what to do. The chemistry of the cell is controlled by proteins called enzymes, which act as catalysts for chemical reactions. And the structure of each enzyme is encoded by an individual piece of DNA, a gene. A chromosome may contain many thousands of genes, but not all of them are directly concerned with the coding of enzyme structure.

The action of the genes in directing the synthesis of enzymes is extremely complex and involves production from the DNA of different forms of the related nucleic acid, RNA. The RNA passes out of the nucleus and takes part in the job of enzyme synthesis on small particles in the cytoplasm called ribosomes. The ribosomes are themselves made of protein, manufactured in the cytoplasm, and RNA which comes from the nucleus. Thus the activity of the nucleus necessitates a continual export of material into the cytoplasm. This export occurs through pores in the envelope which surrounds the nucleus.

6.4 This view of the external surface of a nucleus was produced by freezing the cell, fracturing it in a vacuum, and then coating the specimen with metal. The micrograph is of the metal 'replica'. It shows part of a nucleus

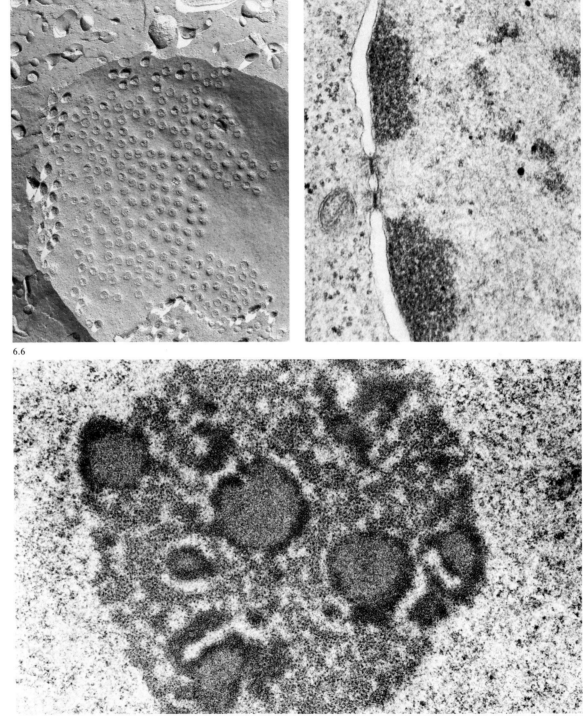

6.4

6.5

6.6

together with surrounding cytoplasm. The nuclear envelope is randomly studded with circular pores; some areas lack pores entirely. At the bottom of the picture, the envelope has split, revealing the inner membrane layer.
TEM, freeze-fracture replica, ×16 000

6.5 In this high-magnification detail of a sectioned cell, the nuclear envelope is seen edge-on, separating the nucleus at

right from the cytoplasm at left. The envelope consists of a pair of membranes, separated by the white space. Two pores are visible, each closed by a dark-staining diaphragm. The large, dark, granular masses hugging the inside of the nuclear envelope are chromatin.
TEM, stained section, ×50 000

6.6 The nucleolus is the region of the nucleus where genes which specify the

structure of ribosomal RNA are active. In this section through a nucleolus, the dark-staining material is RNA which will form into ribosomal particles within the cytoplasm. The very fine granular regions are sites at which RNA synthesis is occurring on the ribosomal RNA genes. The coarser granular material is ribosomal RNA awaiting export to the cytoplasm.
TEM, stained section, ×25 000

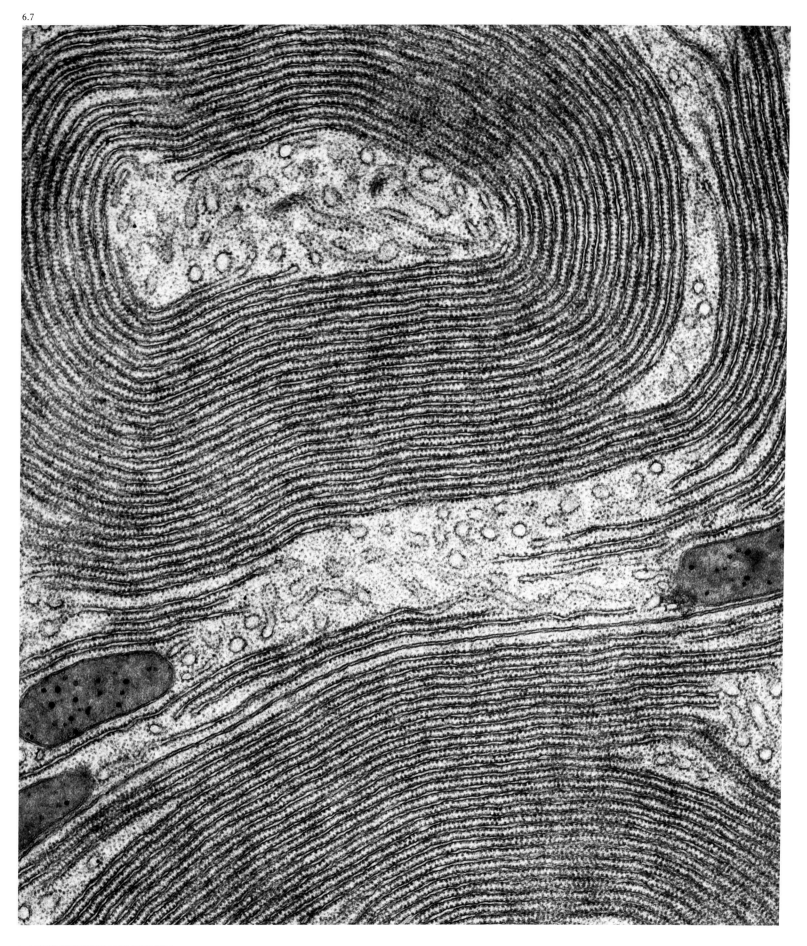

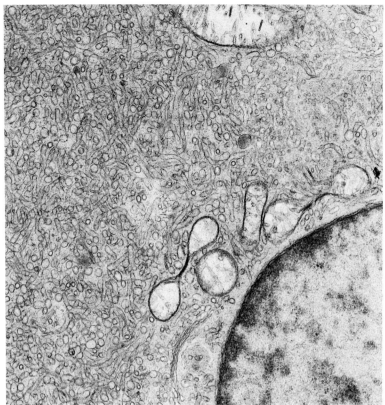

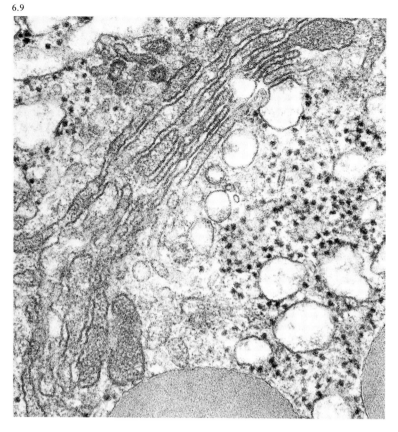

ENDOMEMBRANE SYSTEM

The membrane system of the cytoplasm has two main components – the endoplasmic reticulum and the Golgi apparatus. The endoplasmic reticulum occurs in two forms. *Rough endoplasmic reticulum* consists of extended sheets of paired membranes, and the external surfaces of these membranes are covered with ribosomes. This type of endoplasmic reticulum is found in cells actively engaged in protein synthesis. The chemical process of protein synthesis takes place on the ribosomes, but further processing of the proteins, and their movement to other sites in the cell, can take place within the enclosed space – the lumen – between the paired membranes. Examples of cells which have well-developed rough endoplasmic reticulum are those producing

protein hormones in animals, such as the insulin-synthesising cells of the pancreas, or storage proteins, such as cells in developing leguminous seeds of peas and beans.

The other type of endoplasmic reticulum is called *smooth endoplasmic reticulum*, because it lacks ribosomes. It exists in the form of tubes, or as distended, paired sheets of membrane, and is found in cells which are engaged in the synthesis of lipids, steroids, or other hydrophobic polymers. The synthesis of these materials takes place within the lumen of the endoplasmic reticulum; again, products may move about the cell within this enclosed space.

The Golgi apparatus, named after an Italian count, consists of a series of membranous organelles called dictyosomes. Each dictyosome is a stack of paired membranes, and it acts as a sorting office, packaging the products of the endoplasmic reticulum and directing them to their correct destination. Materials enter the dictyosome at one end of the stack of membranes – called the forming

face. They leave the dictyosome wrapped up in small spherical vesicles which are budded off from the other end of the membrane stack – the maturing face. On their passage through the membrane stack, they may be chemically modified in order to ensure their correct arrival at their ultimate destination.

The Golgi apparatus is particularly well developed in cells which are exporting a chemical product. Thus dictyosomes are common and active in cells producing digestive enzymes in animals, and they are also found in large numbers in plant cells which are growing rapidly and therefore producing large amounts of wall materials. The vesicles fuse with the cell's plasma membrane and discharge their contents into the extracellular space.

6.7 This micrograph is of rough endoplasmic reticulum in a cell from the pancreas of a bat. The paired membranes appear in large whorls, with tiny black ribosomes attached to their outside surfaces. The space within the endoplasmic reticulum

membranes is continuous; it is also separate from the cytoplasm. This cell is engaged in the synthesis of digestive enzymes; the total surface area of endoplasmic reticulum in such cells exceeds 1 square metre for every cubic centimetre of cell volume.
TEM, stained section, ×31 000

6.8 This cell, from the testis of an opossum, is engaged in steroid synthesis. The cytoplasm is packed with smooth endoplasmic reticulum in the form of branching tubes. The large round body at bottom right is the cell nucleus; the other organelles visible are mitochrondria.
TEM, stained section, ×15 000

6.9 This section through a single dictyosome shows the forming face at lower left and the maturing face at top right. The dictyosome consists of a stack of six paired membranes curving diagonally across the picture. Within the lumen of each paired membrane appears a grey, granular deposit – the protein destined for export. The white areas, bounded by a membrane and surrounded by the black dots of ribosomes, are fragments of rough endoplasmic reticulum. The large grey bodies at bottom are droplets of lipid.
TEM, stained section, ×30 000

MITOCHONDRIA

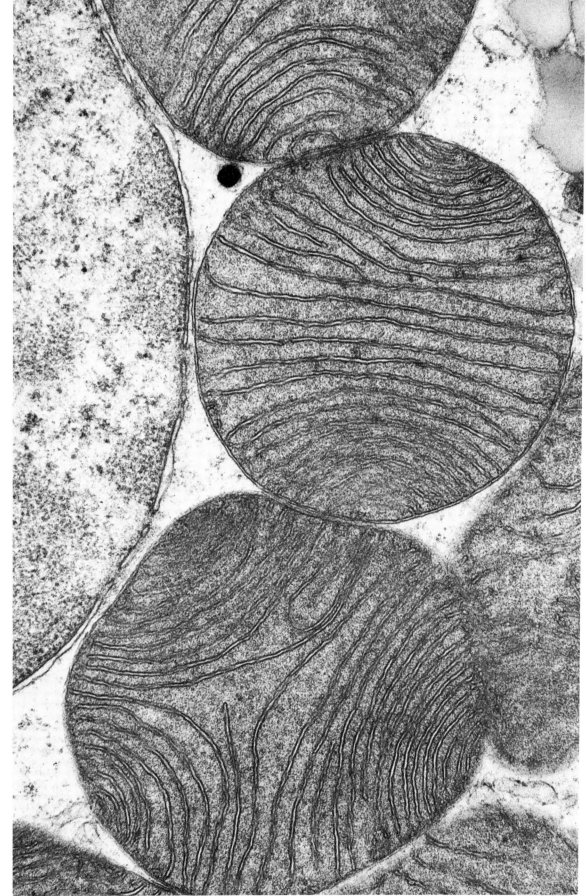

Mitochondria are found in all eukaryotic cells, plant and animal. They are the site of cell respiration – the chemical process which uses molecular oxygen to oxidise sugars and fats in order to produce energy. The energy is stored in the form of a small molecule, adenosine triphosphate or ATP, which is used throughout the rest of the cell to drive chemical reactions such as those involved in forming new proteins, and to power movement within the cell.

Mitochondria are about the same size as bacteria, and they are just visible in the light microscope. Optical examination shows them as small thread-like structures in living cells, continually changing shape and moving in the cytoplasm.

The mitochondrion is bounded by a double-layered membrane. The inner membrane is folded extensively to produce ingrowths called *cristae*. The cristae are where the complex reactions of respiration take place. The mitochondrion's internal fluid, the 'matrix', contains large numbers of ribosomes, together with a small amount of DNA. Using this DNA, the mitochondrion is able to make some of the proteins it requires to carry out respiration. The remainder are imported from the surrounding cytoplasm.

A cell may contain dozens or hundreds of mitochondria. When it divides, its mitochondria are partitioned between the two daughter cells, each receiving roughly half the total. To maintain their numbers through repeated cell divisions, the mitochondria themselves divide. This is accomplished by a process of binary fission similar to the division of bacteria.

In certain environments, mitochondria develop a fixed

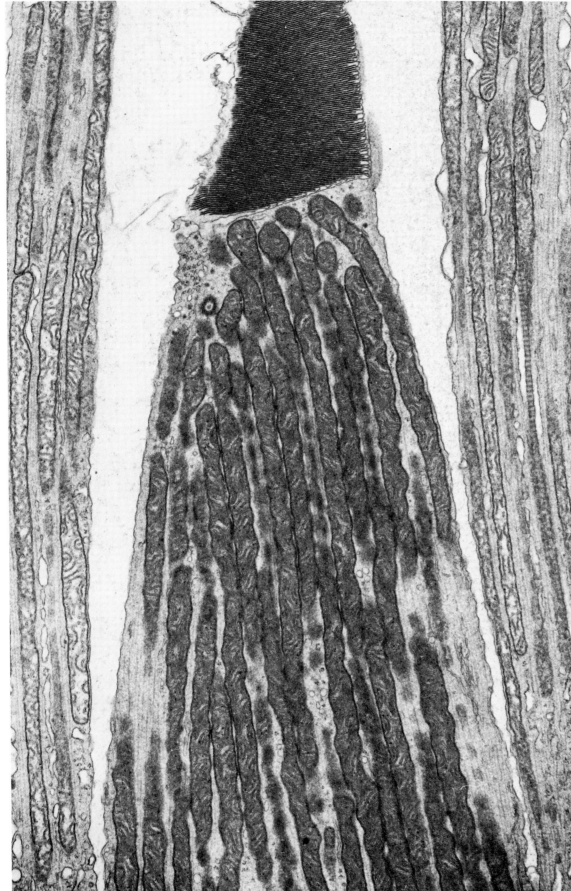

relationship to other cell components. In heart muscle, for example, they are always found close to the muscle fibres, and in sperm they are wrapped round the tail. These arrangements allow ATP produced by respiration to reach its site of action as quickly as possible.

Mitochondria are thought to have arisen by the modification of intracellular bacteria early on in the evolution of eukaryotic cells. They are similar to bacteria in size, they divide like bacteria, and their ribosomes are poisoned by some antibacterial antibiotics. If this idea is correct, it demonstrates an extremely subtle form of symbiosis; mitochondria depend on the rest of the cell for many of their proteins, and the cell depends on the mitochondria for its energy.

6.10 These mitochondria in a brown fat cell from a hibernating bat are particularly large, about 5 micrometres in diameter. Their cristae form narrow bands extending right across the matrix of the organelle. To the left of the picture, part of the cell nucleus is visible. These mitochondria generate the heat necessary to arouse the bat from its hibernation.
TEM, stained section, ×43 000

6.11 The human retina has two types of photoreceptive cells, called rods and cones. This transmission electron micrograph shows a section through a cone cell (centre) and part of two rods (left and right). Both cell types have very elongated mitochondria; those of the cone cell are stained more densely. The black, very fine and closely stacked layers, or lamellae, at the top of the cone cell are the site of the photosensitive pigment. The process of photoreception is poorly understood, but the development of such striking arrays of mitochondria suggests that it requires large amounts of energy.
TEM, stained section, ×3900

CHLOROPLASTS

Chloroplasts are membrane-bound organelles which occur in algae – except the blue-green algae – and in green tissues of all higher plants. They are the sites of photosynthesis, and contain the green pigment chlorophyll in a highly ordered arrangement within their internal membranes. Light falling on the chlorophyll releases electrons which initiate a complex chain of chemical reactions leading to the formation of sugar (sucrose) from carbon dioxide. The sugar may be stored within the chloroplast as starch, or it may be exported throughout the plant via the phloem of the vascular network. Molecular oxygen – the 'breath of life' to ourselves and other animals – is a waste product of photosynthesis.

Chloroplasts are larger than most mitochondria and typically lens-shaped, with a long axis of about 5 micrometres. They are bounded by two membranes; the inner one gives rise to a series of interconnected stacks called *grana*. The interconnections between the grana are known as 'frets'. The fluid within the chloroplast, the *stroma*, contains ribosomes, regions of DNA called nucleoids, and sometimes fat droplets or starch grains. Like mitochondria, chloroplasts are able to make some of their own proteins, but not all; and they divide by binary fission.

Chloroplasts are one member of a family of organelles called plastids, which have different functions depending on the type of cell in which they occur. Photosynthetic cells have chloroplasts, but in storage tissues such as the potato tuber, the plastids exist as amyloplasts, with large starch grains. In carrot roots and the petals of yellow flowers, the plastids are chromoplasts, containing yellow pigment molecules.

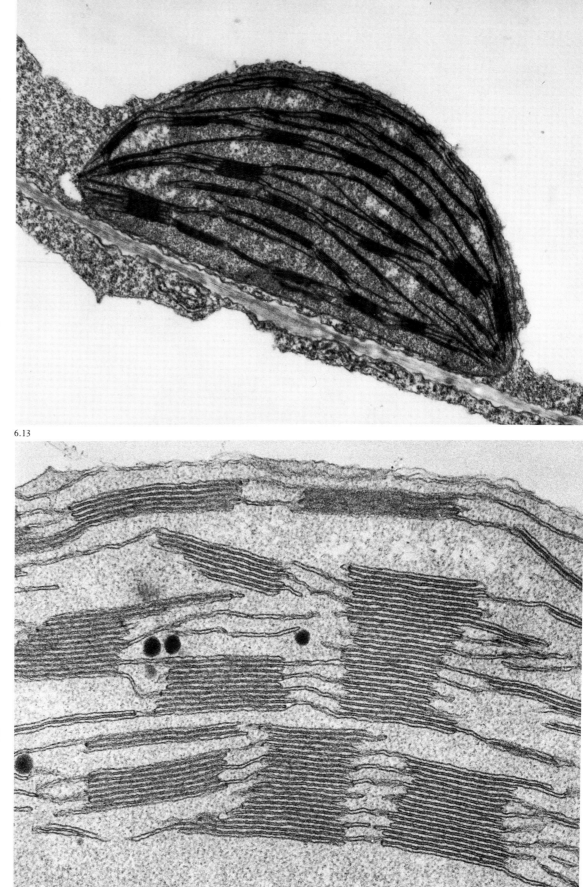

6.13

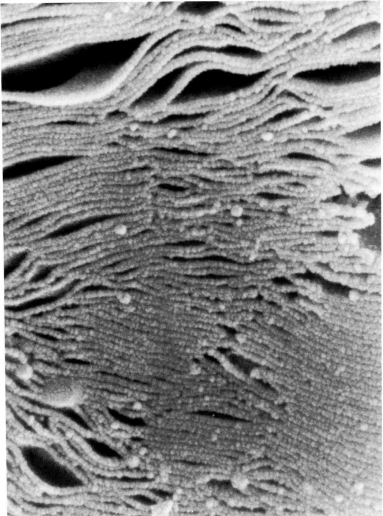

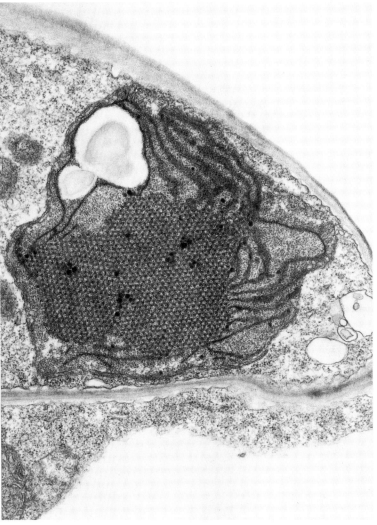

The number of chloroplasts in a cell is variable. Some algae possess only one per cell, whereas the leaves of higher plants usually contain 20–50 per cell. This means that every square millimetre of leaf surface exposes about 500 000 chloroplasts to the light.

As with the origin of mitochondria from bacteria, chloroplasts are thought to have started as intracellular prokaryotic cells, similar to the present-day blue-green algae. Genetic evidence shows that the protein-synthesising machinery of the chloroplast resembles that of prokaryotes.

6.12 In this transmission electron micrograph of a chloroplast in a leaf cell of tobacco, *Nicotiana tabacum*, the lens shape is clearly demonstrated. The grana appear as closely packed membranes connected to each other by a number of single-membrane frets. The paler regions in the stroma are nucleoids, where the chloroplast DNA is situated.
TEM, stained section, ×27 700

6.13 At higher magnification, the stacked structure of the grana is more obvious. The reason for this formation is not clear, since photosynthesis can occur in plants which do not form granal stacks. The black particles in this picture of a leaf chloroplast of maize, *Zea mays*, are fat droplets. They act as a reserve of raw materials for the production of new membranes.
TEM, stained section, ×43 500

6.14 An entirely different view of the inside of the chloroplast is given by the scanning electron microscope. This picture shows the membranes comprising the granal stacks edge-on. The granular appearance is due to the method of preparing the specimen, which is a dried and broken chloroplast from the spotted laurel, *Aucuba japonica*. This is a very high-magnification picture for scanning electron microscopy.
SEM, fractured dried specimen, ×62 500

6.15 Chloroplasts of flowering plants need light for their formation. A seed germinated in darkness produces pale spindly growth, and such a plant is said to be etiolated. The leaves of etiolated plants contain special plastids called etioplasts. The one in this picture is from a seedling of maize, *Zea mays*. Instead of granal stacks, its membranes have formed a crystalline array, with a few frets. Normal chloroplast structure is generated from this crystalline array within a few hours of an etiolated plant being exposed to light. The white body within the etioplast is a starch grain.
TEM, stained section, ×31 500

CYTOSKELETON

The cytoplasm of cells is highly organised. The smaller organelles are in continuous motion, and the chemical products of metabolism are distributed precisely to different sites within the cell. The basis of this organisation is poorly understood, but advances in light microscopy have shown that the cytoplasm is permeated by networks of tubular and filamentous proteins. These three-dimensional arrays of protein molecules are collectively known as the cytoskeleton.

The components of the cytoskeleton fall into two groups. The *microtubules* are rigid protein tubes assembled from subunits. They are thought to act as direction markers within the cell – a sort of cytoplasmic railway. The filamentous components of the cytoskeleton (*microfilaments* and *intermediate filaments*) form three-dimensional networks, and are thought to be directly involved in the movement of organelles, vesicles and membranes, probably by a sliding mechanism similar to that found in muscle fibres.

The arrays can be visualised by staining the protein molecules with antibodies coupled to fluorescent dyes, and viewing the cell in an ultraviolet fluorescence microscope. It is likely that all cells contain a cytoskeleton, but because of technical difficulties associated with the plant cell wall, most work has been carried out on animal cells.

6.16 This picture, of a group of animal cells which have been grown in liquid culture, shows the microtubule component of the cytoskeleton. The cells have been fixed and stained with an antibody which binds to microtubules. The antibody has been chemically attached to the dye fluorescein, which produces an intense green fluorescence in ultraviolet light.

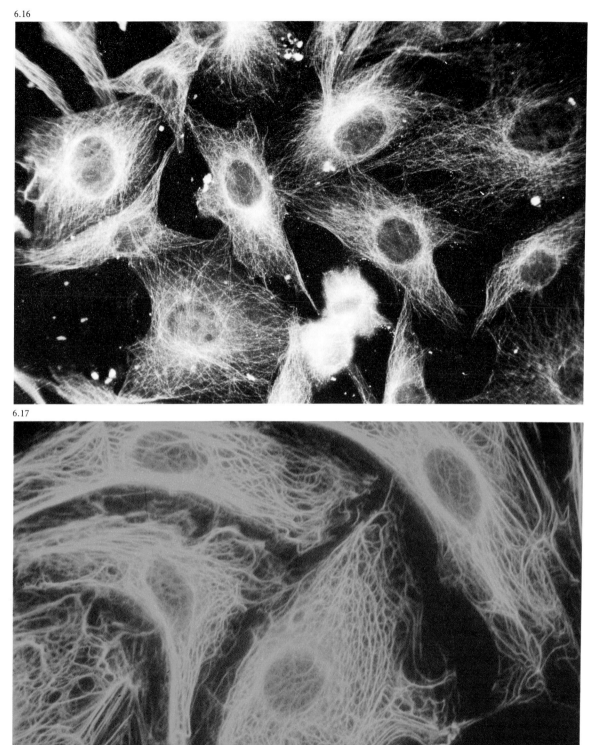

6.16

6.17

The large, apparently empty space in the centre of each cell is the nucleus. The microtubules radiate out from the nucleus to the cell periphery. They are predominantly straight.

LM, ultraviolet fluorescence, anti-tubulin stain, ×200

6.17 This micrograph shows the distribution of an intermediate filament protein called vimentin. The cells are from the kidney epithelium of the kangaroo-rat; they have been grown on a glass cover slip, fixed, and stained with antibody to vimentin,

coupled to fluorescein. Superficially similar to Figure 6.16, this picture shows that intermediate filament proteins are flexible and form networks.

LM, ultraviolet fluorescence, anti-vimentin stain, ×440

SPECIALISED CELLS

Higher organisms contain various cell types of different specialities. What a cell does is determined by the activity of genes within its nucleus. Although each nucleus effectively contains a huge library of information and blueprints, in any particular cell only a few volumes of that library will be acting as instructions.

The way in which cells specialise may be biochemical – the production of a particular metabolite, for example. Or it may involve considerable structural modification; the development of networks of endoplasmic reticulum and the production of chloroplasts are examples already illustrated in this chapter. Cells may specialise in an ability – to sense gravity, for example, or the colour and intensity of light. Whatever their function, they act with other cells to produce a working tissue, such as a liver, or a leaf. It is this ability to act in a coordinated fashion which distinguishes the cells of highly evolved creatures from those of more simple groups.

6.18 Muscle fibres consist of a highly organised arrangement of two types of fibrous protein which slide over one another during muscle contraction. This transmission electron micrograph shows a cross-section of the fibres in the flight muscle of the fly, *Bombylius major*. The black circles arranged in a hexagonal pattern are sectioned myosin filaments. Each one is surrounded by six black dots, which are sections through actin filaments. The two filament types are linked at intervals by cross bridges – visible as the hazy grey halo around the myosin filaments. Muscle contraction involves the breakdown of large amounts of ATP, and so muscle fibres are often closely associated with energy-producing mitochondria.
TEM, stained section, ×31 000

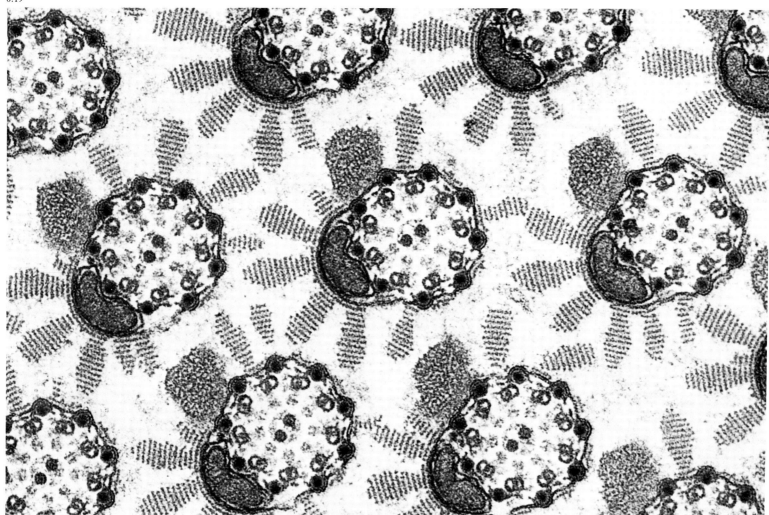

6.19 Sperm cells are specialised for swimming: they are streamlined in shape, and their cytoplasm is reduced to bare essentials. They have no endoplasmic reticulum or Golgi apparatus, and the nucleus is very densely packed with DNA. The cytoplasm contains many mitochondria, which provide the energy for movement. The cell consists of two parts: a head region, containing the nucleus, and a long tail, or flagellum, which has mitochondria wrapped round its base. Movement is accomplished by the rhythmic beating of the flagellum. This picture shows a group of sperm tails in cross-section, during their maturation in the testis of a moth. Each contains an array of microtubules – two dead centre, surrounded by a group of nine microtubule doublets and, beyond those, nine single microtubules pressed against the plasma membrane of the cell. The central pair and the outer nine microtubules are filled with a stained material and appear solid in

this species. The beating of the flagellum is accomplished by the microtubule doublets sliding past one another in a process which is fuelled by ATP. The large granular body within the plasma membrane, at bottom left of each flagellum, is called a 'dense fibre'; its function is unknown. Also unknown is the reason why, in this species, the outer surface of the plasma membrane is decorated with striated projections.
TEM, stained section, ×83 000

6.20 The surface of a cell is usually more or less smooth, but in certain situations cells develop bristle-like extensions called brush borders. These cells are found in tissues where rapid absorption is taking place, such as in the surface of the intestine, and in the kidney. This picture was obtained by rapidly freezing a piece of kidney tissue, then fracturing it in a vacuum and coating the exposed broken surface with platinum. The micrograph is of the platinum replica.

It shows a brush border in the kidney epithelium. Each of the bristles – called *microvilli* – contains bundles of actin filaments. The effect of the brush border is to increase the surface area of the cell about 25 times. The large objects, top right and bottom left, are surface views of cell organelles.
TEM, freeze-fracture replica, ×17 500

6.21 A second type of bristle-like extension to the cell surface is shown here. Each bristle is called a *cilium*, and it contains a bundle of parallel microtubules. The microtubules grow out from 'basal bodies' – seen here in a row lying just inside the cytoplasm, and associated with a dark-staining granular material. Cilia perform a beating movement by sliding adjacent microtubules over one another. The function of the beating motion varies with the cell type. Some protozoa are covered in cilia, and use them to propel themselves through water, or to set up currents that attract food particles. The surface of the

respiratory tract in animals is also ciliated; here the beating of cilia sweeps layers of mucus, dust and dead cells up towards the mouth. Cilia also help to sweep egg cells along the oviduct. Ciliary movement requires energy, and this is supplied by mitochondria, several of which can be seen at bottom left of this picture.
TEM, stained section, ×25 000

6.22 The liver is the organ of the body which processes nutrients from the digestive tract and either stores them or makes them available for use elsewhere. In this micrograph of a liver cell from the slender salamander, *Batrachoseps attenuatus*, the cytoplasm can be seen to contain large numbers of dark-staining crystals. These are made of protein, and have formed within the lumen of the rough endoplasmic reticulum. They probably represent reserve storage material. The cell nucleus, with its nucleolus and peripheral chromatin, is bottom right.
TEM, stained section, ×6250

6.20

6.21

6.22

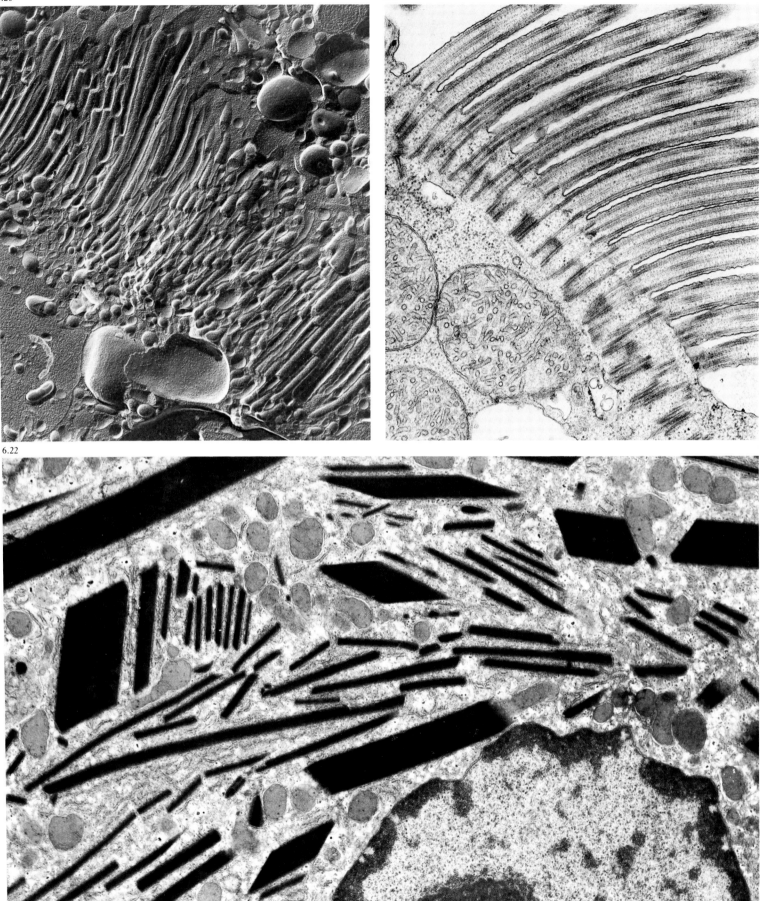

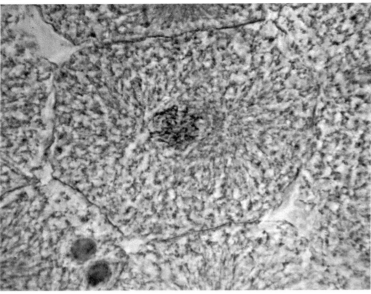
6.23a

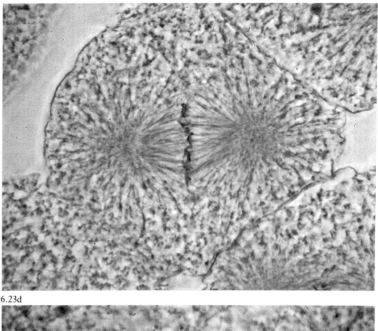
6.23b

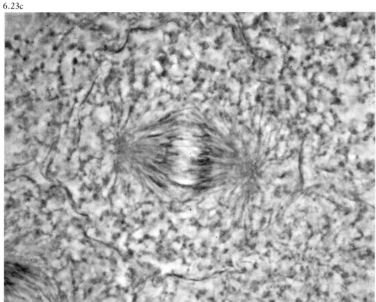
6.23c

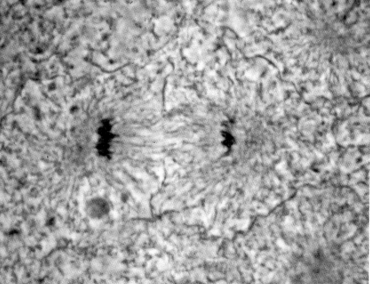
6.23d

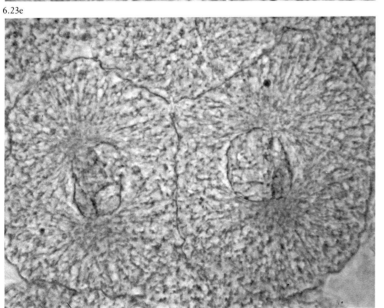
6.23e

MITOSIS

The main genetic material of a cell is contained within the nucleus, where it exists in duplicate as paired chromosomes. The cell division cycle consists of two stages: interphase, when all the nuclear DNA is carefully replicated to produce the duplicate of itself; and mitosis, during which the two pairs of chromosomes are precisely distributed by a mechanical process. The result of

this cycle is that both the new cells formed by cell division contain exactly the same genetic information as the starting cell. Mitosis involves an elaborate series of movements on the part of the chromosomes, mediated by a structure called the mitotic spindle.

6.23 This series of light micrographs illustrates the main phases of mitosis in an animal cell. The whole process usually takes from one to a few hours. Prophase, the beginning of mitosis, is characterised by the condensation of the chromosomes and their appearance as a tangle of dark-stained material in the nucleus (6.23a). Next, the nuclear envelope dissolves, and the mitotic

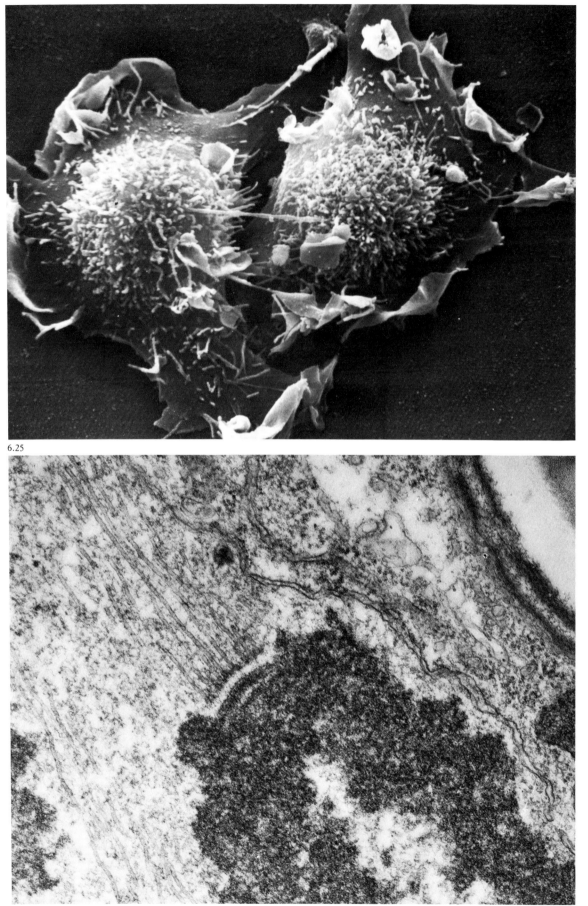

6.25

spindle forms. For a time, the chromosomes make apparently random movements. Eventually, when all the chromosomes are firmly attached to the spindle, they line up at its centre, and movement temporarily ceases (6.23b). This pause is known as metaphase. It may last for a few minutes, or for many hours. At the end of metaphase, the chromosomes split into two equal halves, called chromatids, which move in opposite directions to the poles of the spindle structure. This movement is known as anaphase (6.23c). When the chromatids reach the spindle poles, they group together and begin to reform a nucleus – the stage known as telophase (6.23d). Finally, the two new nuclei are formed, complete with envelopes, and separated by a new cell membrane (6.23e). The nuclei are now in interphase, during which no chromosomes are visible because they are dispersed and do not stain well.
LM, bright field illumination, all ×900

6.24 The final separation of the two daughter cells following division is shown here. The slender thread which joins them is about to be severed. Each daughter cell has a rough surface covered with microvilli in the central region which contains the nucleus; the peripheral regions are smoother and quite thin.
SEM, ×3000

6.25 Through the light microscope, the mitotic spindle appears fibrous; electron microscopy shows that the fibres consist of bundles of microtubules. This picture is of the 'kinetochore' region of a chromatid – the point at which it is attached to the mitotic spindle. The chromatid appears as the dark-stained granular body. The kinetochore is the plate-like region to which several microtubules (the parallel lines) are attached.
TEM, stained section, ×44 000

CHAPTER 7
INORGANIC WORLD

IN the modern world we are surrounded by the results of man's endeavours to improve the raw materials from which he fashions his utensils and playthings. Microscopy has played a preeminent role in the study of inorganic materials and the quest to exploit their physical characteristics. It is difficult to imagine a modern aircraft, capable of carrying hundreds of passengers, without the use of strong light alloys for the structure and tough heat-resistant alloys for the jet engines. It is equally difficult to imagine the development of the electronic chip without the availability of pure silicon crystals. These materials are only manufactured today because of the years of background study by scientists using microscopes.

The subjects of this chapter are inanimate; they are incapable of reproduction. They include metals, alloys, minerals and ceramics. Although they are all different types of materials, their study converges at the atomic level. Solids are characterised by ordered atomic structures regardless of whether the atoms are identical, as in a pure metal, or a collection of different atoms, as in an alloy or compound. The atomic arrangement or crystallography is a characteristic of a particular material. The majority of atoms within a material will be positioned in a similar manner and will be surrounded by an identical arrangement and spacing of neighbouring atoms. Crystallographers call this identical, repeating arrangement the *unit cell*.

Only fourteen types of unit cell can exist. This was proved by the French mathematician Auguste Bravais in 1848, and hence they are known as the Bravais lattices.

They comprise seven basic shapes and seven slight variations. The names of the seven basic shapes, which are all parallelepipeds, are: cubic, tetragonal, hexagonal, trigonal, orthorhombic, monoclinic and triclinic. The number of variations depends on the type. Cubic, for example, has three: *primitive, body-centred* and *face-centred*.

As long as the ordered structure prevails, a material is solid. When it becomes liquid, the atomic bonds are destroyed and the structure breaks up; the atoms are released to roam randomly throughout the liquid. If the liquid is then cooled sufficiently, the ordered structure is remade and the solid state regained.

The resistance of solids to deformation results from the reluctance of atoms to allow disturbance to their neat atomic arrangements. Deformed solids are under *internal* stress as long as their atomic structures remain deformed. Such objects are likely to change shape if they are heated to allow rearrangement of their atoms. The addition of 'rogue' alloying atoms in the atomic structure can create interatomic stresses which improve a material's resistance to *external* stress and so harden and strengthen the alloy in comparison to the base metal. The most common alloy is steel, made by combining two naturally weak materials: iron and carbon. When they are alloyed, the carbon, which is added in very small quantities (typically 0.2–0.6 per cent by weight), takes up residence in the iron's cubic unit structure and profoundly affects its characteristics. Tenfold increases of strength can be brought about, vastly improving the value of the metal as an engineering material.

The constituents of an alloy, mineral, or ceramic are known by a variety of names according to the shape, crystallography, or chemical composition that they possess. *Phase* is the most common of these names and a constituent requires a particular combination of chemistry and crystallography before it can be classed as a particular phase. This may only define it partly; an additional description by shape might be added if a variety of shapes are commonly observed.

The secrets of atomic structure can only just be observed in the transmission electron microscope. At the highest magnifications available, individual atoms can be resolved and their relationship to other atoms studied. But in order to completely characterise an atomic structure, X-ray techniques are also employed. The neat lines of atoms bend X-rays in regular patterns which can be interpreted by the microscopist to reveal not only the type and shape of the atomic arrangement but also its dimensions.

The microscope samples used in this chapter differ vastly from those encountered so far. The delicate sections of the life sciences are supplanted by the more robust but equally elegant preparation techniques of the materials scientist. Abrasives and fine polishing diamonds are used to work the specimens to mirror-quality finishes, or to grind rock samples to wafer-thin transparency. The staining techniques of the life scientist are replaced by etching with chemicals or, in the case of pure ceramics, etching by subjection to temperatures in excess of 1500 degrees centigrade.

Other specimen preparation and decoration techniques are unique to materials science. The use of reflected light enables thin coatings to be set down onto specimen surfaces to decorate phases in metals so that microstructural shapes and patterns can be discerned. Variations of hardness in alloying constituents are exploited in 'relief polishing', which causes hard phases to stand out above the surrounding matrix when prolonged diamond polishing is employed.

The use of reflected light prevails in metals microscopy in order to overcome the inherent opacity of all but the thinnest specimens. Only in petrology – the study of rocks – do microscope techniques bear a resemblance to those of the life sciences, for the petrologist employs transmitted light illumination in a majority of his studies.

7.1 The inner, atomic structure of a material controls the outward appearance of its crystalline form. Although the outer surfaces of a crystal may sometimes look complex, there will always be an exact relationship between the angles and shapes of those faces. The common salt (sodium chloride) shown in the scanning electron micrograph has a cubic unit cell and the faces of the crystal cluster are indeed arranged at right angles to each other. However, the cubic unit cell can also form crystals with the shape of octahedrons or rhombic dodecahedrons. The seven simple Bravais cells can generate 32 crystal forms and further variations increase the number to over 50. The crystalline form gives a vital clue to the identity of a material and the microscopist who studies rocks and minerals regularly will be familiar with the majority of the variations that occur.
SEM, ×400

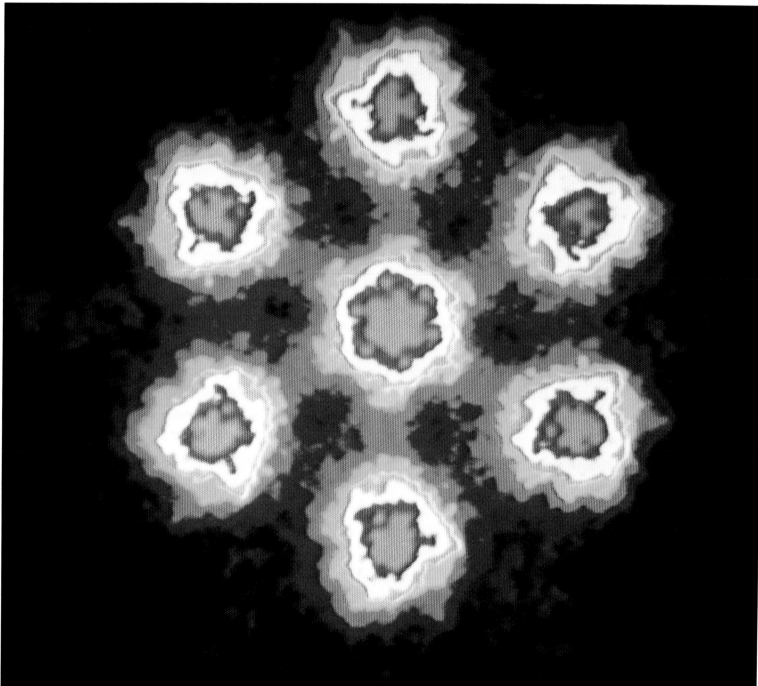

ATOMS

The smallest unit of matter that can be imaged by microscopy today is the atom. The use of high resolution electron microscopy or HREM enables the scientist to study the neat lines and rows of atoms arranged in their unit cells. The world of atomic level microscopy is bathed in hyperbole. Imaging an atom at a magnification of ×100 million is equivalent to observing from Earth the golf ball that Neil Armstrong hit on the Moon. The microscopists at the forefront of high resolution imaging are now trying to read the golf ball's number!

The HREM image is not sharp in the normal photographic sense and the image signal is frequently 'cleaned up' with the aid of a computer and complex mathematical techniques.

7.2 This picture combines scanning transmission electron microscopy (STEM) with false colour computer enhancement. The subject is a uranyl acetate microcrystal and the image shows the uranium atoms arranged in a perfect hexagonal shape around a central atom. Each atom is spaced 0.32 nanometres from its neighbour. The carbon, oxygen and hydrogen atoms, which make up the remainder of the uranyl acetate microcrystal are transparent to the electrons used as the illuminating radiation and do not show. Microscopists exploring the limits of atomic resolution have tended to use the heavy elements, such as lead, gold and uranium because they have the ability to stop electrons and therefore show up well in the electron microscope.
STEM, false colour, ×120 million

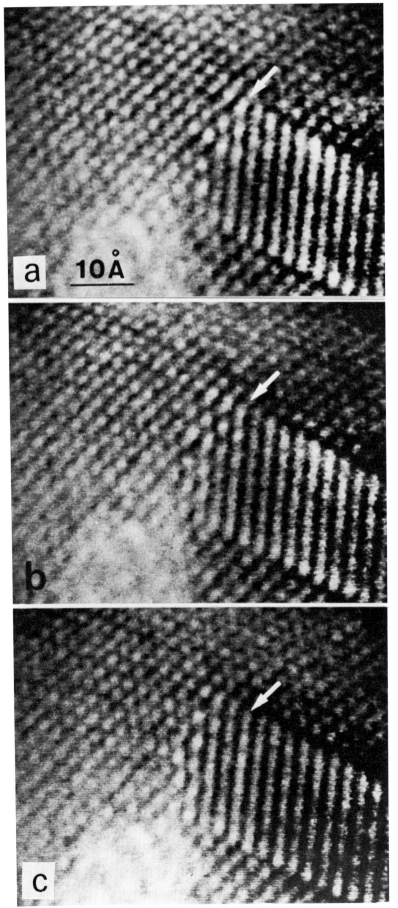

a **10Å**

b

c

7.3 High resolution microscopy reveals the symmetry and order of atoms in a solid material. In nature, this symmetry is usually disturbed when a solid is subjected to stress. Distortions are produced, occurring along planes of weakness where whole blocks of atoms glide over each other in a process known as *slip*. The ability of an atomic lattice to resist slip determines the bulk strength of the material. When multiple blocks of atoms move along parallel slip planes, a process known as twinning occurs. The meeting points of slip planes often contain holes or vacancies where atoms are missing from the regular array. These vacancies are a major influence on a material's properties; they are vital, for instance, to the working of transistors and silicon chips. It is possible to observe the movements of atoms when slip occurs. These micrographs were exposed within 0.07 seconds of each other. The arrow shows the advance of a twin – the band of slanted atomic lattices – in a crystal of gold. The movement of the twin has been induced by both localised atomic forces and the bombardment of the microscope's imaging electrons. The 10 angstrom (1 nanometre) bar in the top micrograph indicates the scale. HREM, ×17 million

7.4 The pattern displayed here is a convergent beam diffraction image, also known as a Kirkuchi pattern after its Japanese inventor. These patterns are uniquely characteristic of a material and may be used to identify the smallest of particles in an alloy. The patterns are created in the transmission electron microscope by focusing a convergent beam of electrons onto the specimen and then photographing the back focal plane of the objective lens, rather than the conventional image. The back focal plane contains the 'diffraction image', which is created by the regular atomic planes within a particle affecting the passage of electrons through it. This deflects the electrons in a highly regular and ordered way which gives rise to a 'fingerprint' which can be interpreted by the microscopist. The example is from a particle of gamma phase in the refractory super-alloy 'Astralloy'. The gamma phase is responsible for Astralloy's high-temperature properties. All the lines of the image, their number, relative angles and disposition confirm that the gamma phase is present in the specimen and that the atoms are arranged in a cubic unit cell. It is not normal to quote the magnification of a diffraction image as it has no meaning. TEM, convergent beam diffraction

DISLOCATIONS & GRAIN STRUCTURES

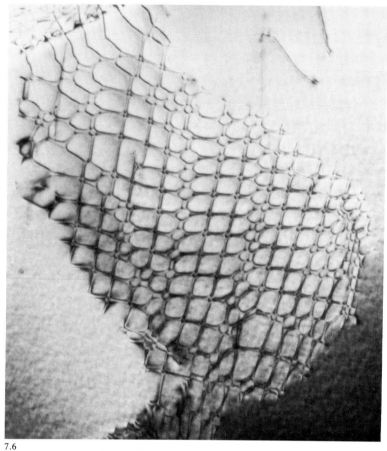

When the scale of microscopic examination is expanded from the atomic, the simple world of ordered structures becomes more complicated. Single atomic defects like those shown in Figure 7.3 begin to collect into groups, forming 'dislocations'. Dislocations are large numbers of defects arranged in particular ways within an atomic structure and crossing many unit cells. They take a variety of forms and some reach sizes which can be observed in the optical microscope or, more rarely, with the naked eye. Dislocations can also move bodily through a solid, sometimes looping and tangling like a length of cotton shaken in a bottle of water. In the transmission electron microscope, dislocations reveal themselves as dark lines.

Many materials are made up of *grains*. Each grain has its own crystallographic orientation, although adjacent grains may have a different orientation. Grain growth can occur in a variety of ways, but the most common is from a cooling liquid. The ordered structure is preferred by atoms changing from the liquid to the solid phase because this is the most efficient way to distribute the interatomic forces. Atoms residing in the correct position in the lattice are in the lowest energy state and so they are less likely to move from these positions. The three-dimensional surfaces which enclose grains are called grain boundaries and are of considerable interest to the materials scientist. Impurities are pushed ahead of a solidifying material and are eventually trapped in the grain boundaries when adjacent grains collide. Strength is often improved by grain boundary effects, but corrosion resistance is usually reduced.

7.5 This tiny piece of a mechanically rolled sheet of a nickel-aluminium alloy displays a network of dislocations. They have piled up along the grain boundary just visible in the top left corner of the micrograph. The development of the dislocations occurred when the cold-rolled sheet was reheated or *annealed*. The density of dislocations in such a material can be as many as 500 000 million in just 1 square inch.
TEM, ×65 000

7.6 Provided the specimen has been suitably prepared and a sufficient viewing magnification is chosen, most materials will reveal a grain structure. The exceptions are the glasses, which have a special type of atomic structure. The alpha alumina ceramic in this scanning electron micrograph has a fairly typical grain structure, revealed by the network of lines. Two grain boundary impurity phases can be seen: a large one just above centre and a much smaller one at bottom right. They are beta alumina phases which contain sodium oxide, a chemical not present in the alpha alumina that makes up over 97 per cent of the material. The holes are areas of porosity, or voids, which have not been eliminated during the ceramic sintering process.
SEM, polished section, thermal etch, ×2000

7.7 This light micrograph of common brass shows that the grain structure of a metal alloy is quite similar to that of a ceramic such as alpha alumina. The specimen has been polished and then chemically etched. Grain boundaries are more readily attacked by the etchant and show up as fine lines around the more or less polygonal grains. The straight, parallel sets of lines running across the polygons are twins which show different shades of colour to the rest of the grain because of the use of polarised light.
LM, polarised light, polished and etched section, ×500

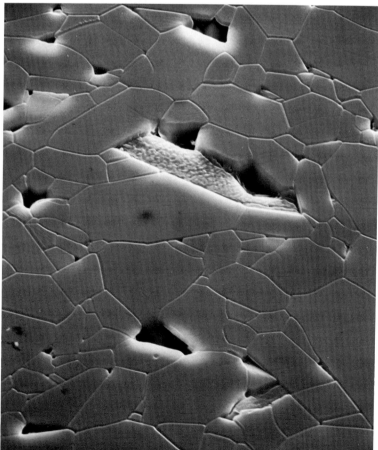

HEAT-TREATMENT STRUCTURES

Some materials possess the useful property of changing atomic structure according to how they are chemically alloyed or heat-treated. The ability to heat-treat steel has made it the most important alloy ever devised. Steel is made by the addition of small quantities of carbon to iron. Below red heat, the iron is arranged in the body-centred cubic or *ferrite* phase. The corners and centre of the cube are occupied by atoms of iron, with the carbon atoms arranged 'interstitially' between the iron. Above red heat, the centres of the cube's faces also contain iron atoms. This is the face-centred cubic or *austenite* phase. Very rapid cooling of red-hot steel will prevent the reformation of ferrite and form a brittle transformation product, *martensite*. Careful rewarming of the steel to control the breakdown of martensite will produce a steel of optimum properties – balancing the hard but brittle martensite with the tougher ferrite. Transformations like this can be observed directly if heat-treatment is carried out on specimens while they are in the microscope.

7.8 This pair of micrographs shows the transformation of austenite to martensite. The austenite (left) was held at 1070 degrees centigrade. It was then cooled to 850 degrees centigrade in 8 seconds so that it transformed into platelets of martensite (right). The picture was taken with the photo-emission electron microscope, which forms its image from the light given off when a specimen is bombarded with electrons. It can also view hot specimens.
Photo-emission electron microscope, ×870

7.9 This micrograph shows typical effects of rapid heat-treatment. The specimen is a magnesium oxide ceramic. It has been heat-treated with a laser beam, and three structural morphologies are shown. The smooth grains at bottom have a well-sintered microstructure – the powder from which the ceramic is made has compacted together at a temperature below its melting point. Above the smooth grains, the ceramic has been melted by the laser beam. The granular region consists of small crystals that have grown only a limited amount because of rapid cooling. At the top, in contrast, more prolonged melting followed by less rapid cooling has produced grains with a 'fir tree' or dendrite structure.
SEM, ×750

7.9

7.8

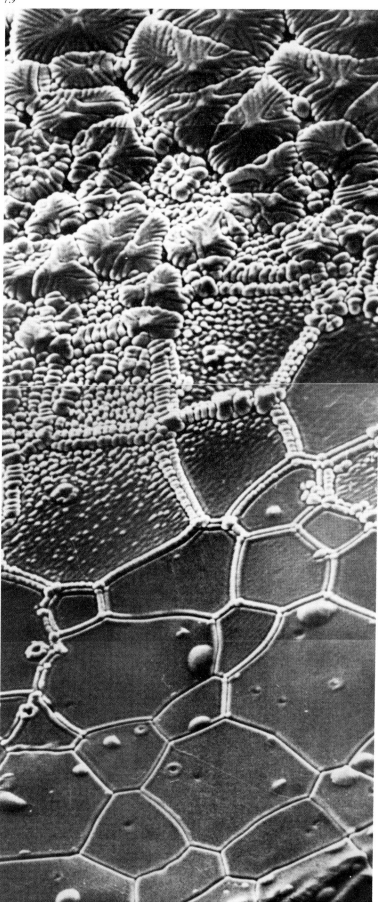

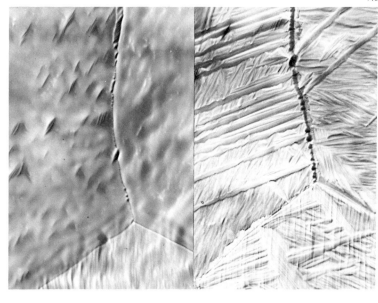

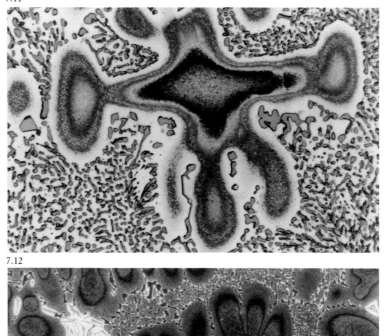

DENDRITIC STRUCTURES

The fairly regular polygons in the lower part of the previous micrograph are the simplest of a whole variety of grain morphologies. The type of grain depends on both the material itself and the way it has been made. When materials are heat-treated, changes occur in their atomic and grain structures and further changes occur on cooling. The dendritic, or tree-like, structure is typical of materials made by casting – solidified in a mold from the molten, liquid state. The process of solidification is both complex and dynamic. The atoms moving from the liquid phase to join the ordered crystal structures of the solid phase prefer to attach themselves to particular atomic planes. Solidifying atoms also emit heat and this aids the growth of 'fingers' of solid within the melt. These primary dendrites rapidly send out secondary arms at specific angles and then tertiary arms at other specific angles until tree-like structures are formed. Solidification is complete when the liquid between each 'tree' finally becomes solid. In a pure material, many dendrites commence growth simultaneously and collide with each other. The boundaries between them form the grain boundaries seen on previous pages.

7.10 These dendrites of an aluminium–titanium alloy (Al₃Ti) were grown under controlled conditions and then exposed by acid in order to study the angular relationships of crystal lattices and dendrite arms. Several primary dendrites have grown outwards from the left-hand edge of the picture. The secondary dendrites that grow upon them are notable for being exactly parallel to their neighbouring primaries.
SEM, acid extracted, ×100

7.11 When an impure liquid (or an alloy of several metals) is cooled, the dendrites are formed from the phase which solidifies at the highest temperature. The remaining melt, depleted in the material that formed the dendrite, will solidify into the interdendritic spacing and thicken the arms of the dendrite. The light micrograph shows a section through an inhomogeneous dendrite formed in this way in an aluminium–silicon alloy used in die-casting. Colour etching has revealed the dark brown, diamond-shaped centre of the dendrite. Around this core are several 'skins', coloured olive green, light brown, and yellow-white, which form side arms to the left and right of the central diamond and lobes above and below it. These skins are formed from increasingly depleted variants of the material that formed the core. Once the dendrite was completed, the remaining material solidified as the eutectic grey and white structure which fills the rest of the field of view. A eutectic is a material which solidifies *en masse* and with the same overall composition at a given temperature; as a result, eutectics have a characteristic 'speckled' structure.
LM, polished section, Weck colour etch, ×100

7.12 This alloy is similar to the one in Figure 7.11 except that magnesium and iron have also been added. The red-brown dendrites are enveloped by a dark blue phase. The interdendritic eutectic phase now has two variants, a predominantly white aluminium–silicon–iron eutectic and a predominantly blue silicon–aluminium eutectic.
LM, polished section, Weck colour etch, ×100

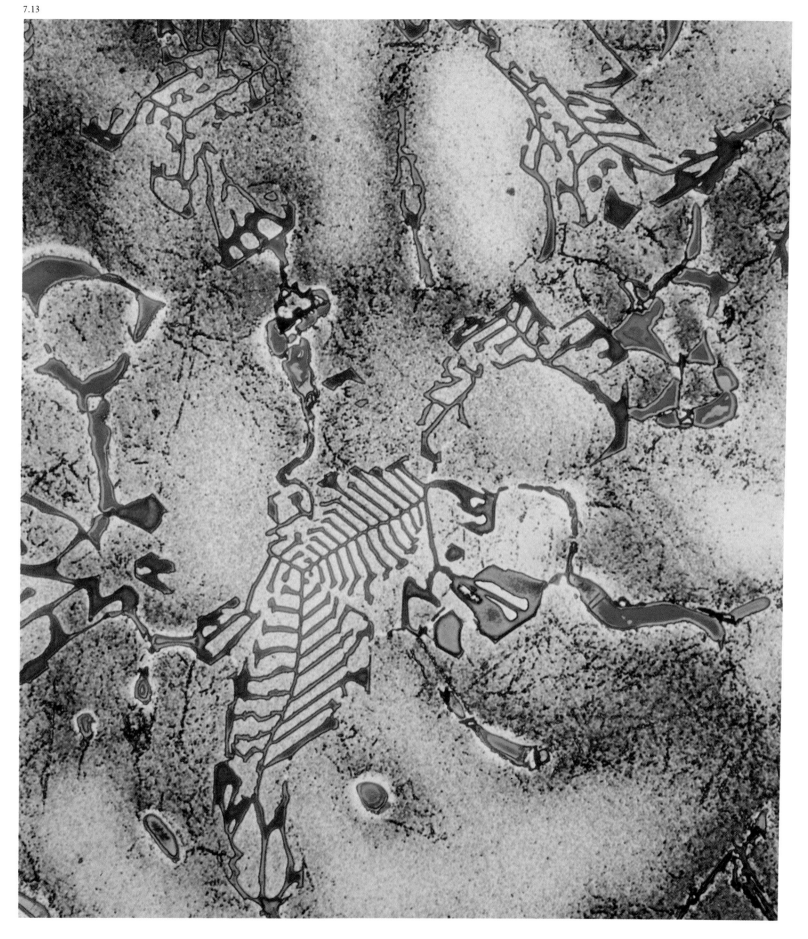

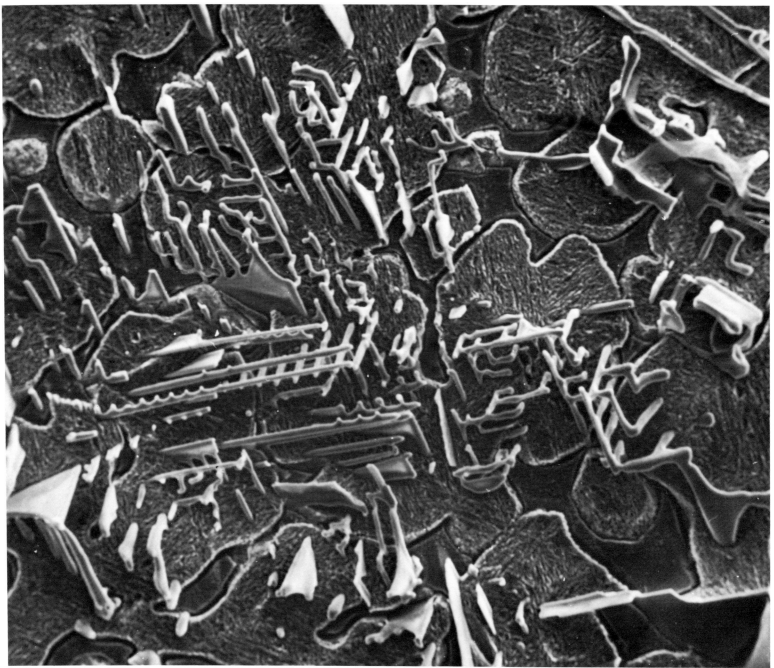

7.13 A large variety of aluminium alloys have been devised to meet the demands of industry. The aluminium–silicon alloy described on the previous page is a binary alloy of eutectic composition – aluminium with approximately 11 per cent silicon. In order to control the size of the dendrites, which can reduce strength if they are too large, the casting melt can be modified by the addition of small amounts of sodium metal. Although this controls the metallurgical quality of the actual casting, it is not possible to remelt the metal remaining in the ladle or any scrap from the molding process. Reducing the silicon and replacing it with copper removes the need to modify the melt with sodium and also improves the alloy strength. Other elements can also be added to improve the castability of the alloy and to effect microstructural changes. These extra alloying elements form complex and numerous phases. The aptly named 'Chinese script' structures in this light micrograph consist of dendrites of aluminium–copper–iron–manganese surrounded by the primary or host dendrites of aluminium–silicon alloy. The spaces between the primary dendrite arms are filled in places by an intermetallic chemical, copper aluminide. These are the small magenta areas with green rims. There is an overall variation of brown shading on a white background across the field of view. After polishing, the sample was immersed in a chemical colouring etchant. This solution deposited the shades of brown according to variations in the chemical composition of the sample – an effect known as 'coring'.

LM, polished section, ammonium molybdate colour etch, ×50

7.14 The complicated shapes that can develop in dendrites are well illustrated in this scanning electron micrograph. The sample is high-speed steel of the type which might be used for the cutting tool of a metal-working machine. The ladder-shaped dendrites have been exposed by prolonged acid etching and consist of metal carbides. The form of the dendrites, and the way their growth has influenced the microstructure around them, results from the use of niobium as an alloying element in the steel.

SEM, deep etch, ×2000

CRYSTAL STRUCTURES

The underlying atomic order that gives crystalline solids their symmetry is only visible in the electron microscope. But the resulting crystal shape is often large enough to be seen in the light microscope or by the naked eye. These large crystals are more in keeping with the layman's idea of what a crystal looks like.

When crystallisation takes place in a narrow space such as the gap between a microscope slide and a cover slip, illumination of the specimen by polarised light produces some of the most vibrant imagery of which microscopy is capable. Although the normal three-dimensional crystal is unable to develop in these circumstances, the underlying atomic symmetry is still present and affects the light waves. The components of white polarised light are diffracted differently so as to give rise to pure spectral colours which give these images their brilliant clarity.

Polarised light observation and photography of crystals is a favourite occupation of many amateur microscopists. For one thing, it is very easy. With the exceptions of glass and amorphous substances such as soot, almost every material crystallises. Many commonly available elements and chemicals can be made to crystallise on a microscope slide either by melting them and then allowing them to cool, or by dissolving them and then allowing a droplet of the solution to evaporate.

7.15 This very pure sulphur was allowed to solidify between a microscope slide and cover slip. Only a 'microscopic' amount of sulphur was used to fill the gap – less than 10 micrometres wide – between the two pieces of glass. The slide was heated until the sulphur melted and then allowed to cool. The cooling caused

spontaneous formation of crystals, which grew out in all directions until they collided, producing irregular grain boundaries. Further cooling produced 'microfractures', which are seen as the series of fine, almost parallel dark lines. The small black dots are voids.
LM, cross-polarised light, ×120

7.16 The crystallisation of chemicals can produce dendritic structures similar to those in solidifying metals and alloys. Although the substances involved are vastly different, the feathery secondary dendrites in these crystals of the pain-killing drug Distalgesic are almost identical to those of the aluminium–titanium alloy in Figure 7.10.
LM, cross-polarised light, ×45

7.17 This scanning electron micrograph of 'fur' from an electric kettle could easily be confused with a horticultural subject. Kettle fur consists of needles of calcium sulphate which precipitate out of hard water into regular crystallographic shapes. The geologist knows calcium sulphate as anhydrite. Its monoclinic crystallographic lattice and flower-like clumps of needles are also the most common form in nature and it is from rocks bearing anhydrite that water derives its hardness. We mine approximately 10 million tons of anhydrite (and its hydrate – gypsum) per annum for building materials.
SEM, ×500

7.18–7.19 These two micrographs are both of vitamin C, a substance which shares the monoclinic crystallography of anhydrite. The scanning electron micrograph could easily be a close-up of the kettle fur in Figure 7.17, since it demonstrates a similar crystal habit. The light micrograph, on the other hand, is quite different in its appearance and perfectly illustrates the effect of constraining the crystal growth to a thin section. This could be achieved in the same way as the sulphur in Figure 7.15 or by allowing a small drop of vitamin C solution to evaporate on the microscope slide.

The cruciform 'interference figure' which appears in the centre of the 'eye' in the light micrograph is a typical polarised light feature and results from the polariser and analyser being at right angles to each other in the optical train of the microscope. The interference figure is normally observed with the eyepiece of the microscope removed; studying its movement as the sample is rotated enables the microscopist to determine which type of crystallographic group the sample belongs to. As with the sulphur in Figure 7.15, a certain amount of microfracturing has occurred. In this instance it is the stress produced as the material dried which has caused the sample to crack. The direction of the stress was at right angles to the fractures and hence was acting almost at right angles to the grain boundary which splits the picture through the middle and deflects around the 'eye'. The jagged 'mountain peaks' running down both sides of the picture are incompletely formed dendrites, which stopped growing when all the available water had evaporated.

7.18 SEM, ×80
7.19 LM, cross-polarised light, ×150

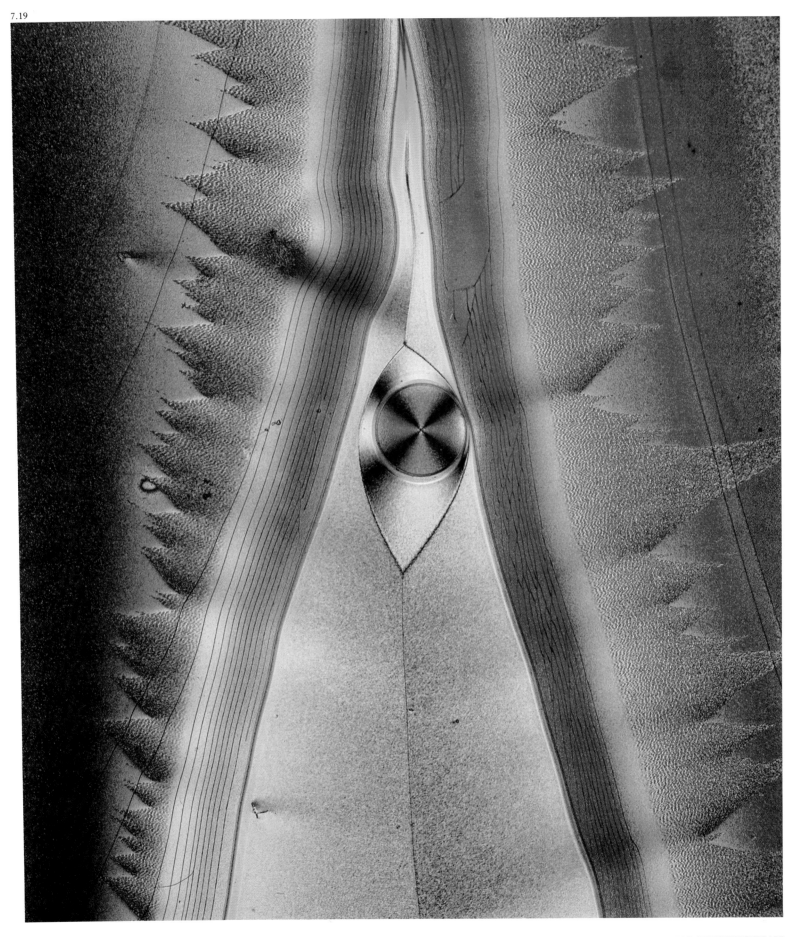

PETROLOGY

The study of rocks is a branch of the geological sciences and is known as petrology. In some ways petrological microscopy bridges the gap between life science and material science microscopy. The petrologist employs the thin section of the life scientist but shares all the concepts of crystallography, solidification and phase with the metallurgist. The petrological microscope is a specialised instrument which has sophisticated polarising facilities and an accurately rotatable stage. Specimens of rock to be examined are glued to glass microscope slides and then ground down until they are 30 micrometres thick. At this thickness, the minerals within the rock are mainly transparent, but when examined in cross-polarised light they show a variety of colours and shades (including grey). Knowledge of these colours and how they change on rotation of the sample enables the petrologist to determine which minerals are present in the specimen.

The petrologist is the historian amongst microscopists – his subjects are generally as old as the Earth. The Earth's core contains a molten mix of rock-forming chemicals called magma which is periodically ejected through the Earth's solid outer crust by volcanic processes. The solidified magma forms *igneous* rocks, the nature of which depends on the composition of the original magma and its thermal history during the process of ejection and immediately afterwards.

7.20 Gabbro is an igneous rock which contains the minerals olivine and plagioclase feldspar. The former is recognized in the microscope by its multitude of randomly oriented fractures, the latter by the multiple twinning of its lath-shaped crystals. In this polarised light micrograph, the twins are seen in a variety of colours because of the use of a sensitive 'tint plate' in the optical train of the microscope.
LM, cross-polarised light with tint plate, ×36

7.21 The upper four pictures on the right-hand page are also igneous rocks. First is a Cornish granite, which has a more complicated mineral assemblage than gabbro. Quartz, orthoclase and plagioclase feldspars, biotite and chlorite are all present. The variously shaded grey crystals are quartz; the twinned crystals are plagioclase feldspar. The mineral biotite shows as a variety of bright interference colours. The mid-grey crystals intruded with veins of a white mineral are orthoclase hosts with albite intrusions.
LM, cross-polarised light, ×7

7.22 This sample is a porphyritic basalt from Mont Dore, France. It displays a quite different texture to granite, although it shares one common mineral – plagioclase feldspar. The feldspar shows clearly as long, 'zebra-striped' twins in a very fine basaltic matrix. Large crystals in a fine-grained matrix are known as *phenocrysts*. The brightly coloured phenocrysts in the specimen are pyroxene and olivine. A feature of this type of rock is that the fine-grained matrix shares the same chemical composition as that of the phenocrysts. The actual size of the phenocrysts depends on the rate at which they were cooled out of the magma.
LM, cross-polarised light, ×7

7.23 This porphyritic rock from Aberdeen, Scotland, displays an even greater difference between its fine matrix and large phenocrysts. This is due to slow cooling during the time when the phenocrysts of clear, grey quartz and turbid, orthoclase feldspar were growing, followed by rapid solidification of the matrix as the volcanic lava was ejected onto the Earth's surface. The large twin towards upper right is a Carlsbad twin, a simpler variation of the multiple twins of plagioclase feldspar.
LM, cross-polarised light, ×7

7.24 Cyclic variations in composition and texture can occur within a single phenocryst in a process known as *crystal zoning*. The highly coloured phenocrysts in this specimen of plagioclase feldspar are all zoned, but the blue and green hexagon at lower right is the most clearly marked. The zoning has occurred because of the changing chemical composition of the molten magma as the phenocrysts grew and floated towards the Earth's surface prior to solidification. The solidification stopped further changes and locked all the variations in place.
LM, cross-polarised light, ×7

7.25–7.26 The deposition of igneous rock is followed in the 'rock cycle' by weathering. The agents of weathering are rain, wind and temperature, which combine to fragment the rock and transport it to lower ground. Materials transported from their source eventually settle as sediments and form *sedimentary rocks*. If sedimentary rocks are buried, they are subjected to heat and pressure and become *metamorphic* rocks. These can be exposed and weathered again, or they may be fully melted and then ejected once more as volcanic rocks to complete the rock cycle.

7.21

7.22

7.23

7.24

7.25

7.26

Figures 7.25 and 7.26 are sedimentary and metamorphic specimens respectively. The former is a 'Greywacke', composed of variously sized fragments in a generally fine matrix. This is typical of deposited materials which have not been transported very far from their source and have had little opportunity to become smooth and rounded. The dark, oval grain at right of centre consists of a multitude of sutured quartz grains and will have been weathered from quartzite, a metamorphic rock.

Figure 7.26 is a sample of quartzite, which is also known as 'Arkose'. The specimen, from Ord, Isle of Skye, consists of more evenly sized and rounded crystals than the Greywacke. This is evidence of protracted transport, which has both smoothed and sorted the grains.

LMs, cross-polarised light, ×7

DIAGENESIS

The process of reforming rocks from weathered fragments is called *diagenesis* when it takes place close to the Earth's surface at low temperatures and pressures. The micrographs on this page show diagenesis in a specimen of sandstone which is in the process of becoming 'solid' or *lithifying*.

7.27–7.28 This pair of light micrographs shows the same thin section in unpolarised (left) and polarised light. The polarised light not only changes the colours, but makes visible structures which cannot be distinguished in the unpolarised image. The sandstone consists of quartz fragments of three types, one above the other. The green in Figure 7.27 is epoxy resin, injected into the porous rock to enable the thin section to be prepared. The upper quartz grain consists of an egg-shaped core, delineated by a brown haematite skin. Its rounded nature tells us that the particle has been shaped and polished by transportation. Around the core, quartz has deposited chemically to produce a crystal-shaped 'overgrowth' with flat faces. The fact that in the polarised light picture the grain's core and overgrowth have the same grey tone indicates that the overgrowth is in perfect alignment with the core. The middle quartz grain consists of 'sutured' subgrains aligned north–south and crossed by the birefringent accessory mineral, muscovite, which appears highly coloured in Figure 7.28. This grain has had a different history to the first grain; it is *schistose* quartz, and has been squeezed during a metamorphic rock-forming process. The bottom grain is of a third type. Although it looks more or less homogeneous in the unpolarised light picture, Figure 7.28 shows clearly that it consists of subgrains that have been joined in random crystallographic orientations, creating quite different colours in the polarised light.

LMs, unpolarised and cross-polarised light, ×150

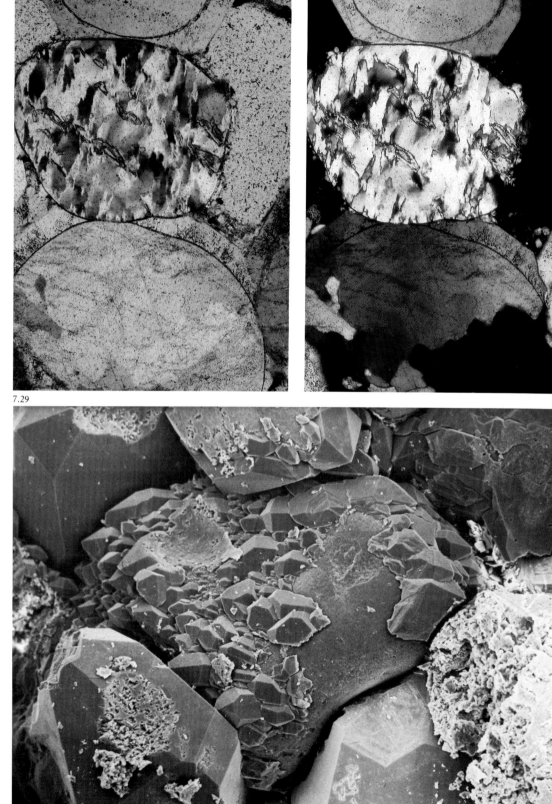

7.27

7.28

7.29

7.29 This scanning electron micrograph shows a part of the same rock from which the thin section above was made. The grain at centre is almost covered with small, angular, crystal overgrowths. This ragged texture is characteristic of partly lithified rock. Such rocks are commercially important because the partly filled voids are capable of holding oil or gas.

SEM, ×100

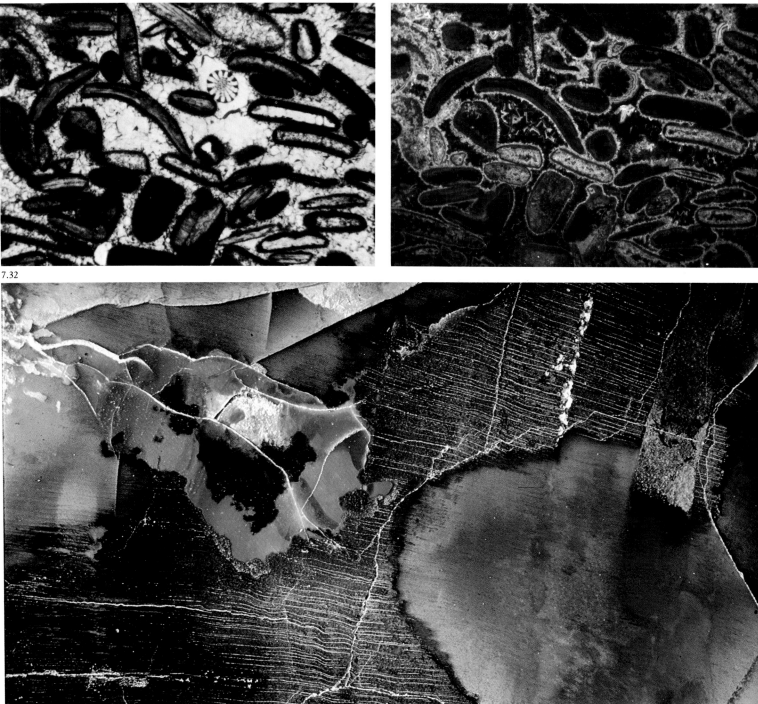

7.30–7.31 This sedimentary limestone demonstrates the use of cathodoluminescence to reveal the innermost details of a rock structure. The rock consists mainly of sea-shell fragments compacted together and then cemented with calcite. Figure 7.30 is a light micrograph made in plane-polarised light. This reveals very little substructure in between the well-defined shell fragments. While it is in position on the stage of a special type of light microscope, the sample can be surrounded by a vacuum chamber and bombarded with high-energy electrons. This causes the calcite cement to luminesce, as in Figure 7.31. The sample absorbs short wavelength radiation (electrons) and emits longer wavelength visible light (in this instance orange). The additional detail produced is particularly evident in the centre of the picture where the fine layers of cement can be resolved. The wheel-shaped sediment is the spine of an Echinoid, which suggests that the rocks are from the Mesozoic or Cenozoic eras. This makes them up to 200 million years old, which is very young in geological timescales.

7.30: LM, plane-polarised light, magnification unknown
7.31: LM, cathodoluminescence, magnification unknown

7.32 The geologist frequently deals with vegetable matter that has been incorporated into the structure of rocks. Typical of this process is the petrified wood shown here. The cellular structure of the wood (brown) has been invaded and replaced by silica, which has deposited from solution to form a rock replica of the original wood. Other minerals surround the wood, the bright red of agate being the most conspicuous.

LM, magnification unknown

FERROUS METALS

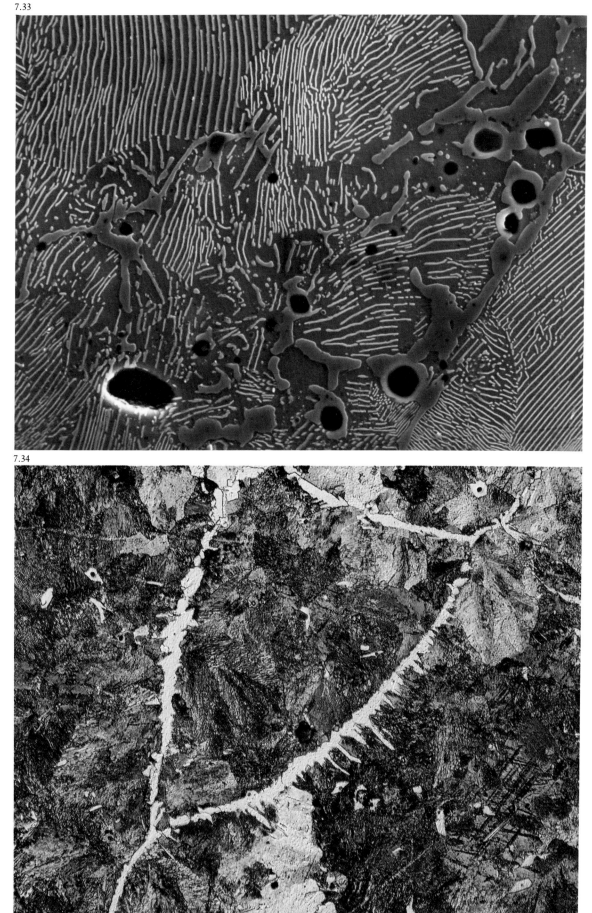

Metals are the most valuable asset of modern society apart from food and water. The fact that two ages of human development, the Bronze and Iron Ages, are named after metals is a testament to their importance. The metals are broadly split into those containing iron – the *ferrous* metals – and those that contain different materials – the *non-ferrous* metals. Steels and cast irons represent the bulk of the former. The addition of carbon to iron makes steel, as described on page 129. As little as 0.05 per cent carbon will begin to affect the properties of iron.

7.33 This micrograph shows a 0.87 per cent carbon steel with a *pearlitic* microstructure of thin platelets. Such a steel is about 10 times stronger than pure iron and a variety of heat-treatment processes could be employed to tailor the physical properties to those required by a design engineer. The platelets, which appear as the bright strips because they are being viewed end-on, consist of alternating sheets of iron carbide or *cementite* and pure iron or ferrite. The platelet structure is formed by the carbon being rejected from the atomic structure as the steel cools.
SEM, polished section, Nital etch, ×3000

7.34 When the carbon content of steel is less than 0.87 per cent, a mixed microstructure is produced consisting of separate areas of pearlite and ferrite. The resulting steel is more ductile, making it specially useful for those applications which require complicated shaping. This example is a 0.4 per cent carbon steel. The yellow triangular feature is a ferrite grain boundary which at an earlier stage of heat-treatment was austenite and has been 'frozen' into the microstructure during the rapid cooling. It has a herringbone or 'sawtooth' structure. The pearlite is visible as a mixture of greens and blues.
LM, Nomarski DIC, polished section, Nital etch, ×100

7.35

7.36

7.35 Iron can hold only so much carbon *within* its atomic structure. Once the carbon level increases beyond 4 per cent, discrete flakes or particles of carbon are formed and the steel becomes classed as *cast iron*. Cast iron is very suitable for applications requiring bulk. The presence of high quantities of carbon also makes the surface of castings hard and durable. This is a 'grey cast iron' which contains its excess carbon as graphite flakes. The flakes aid damping against vibration, a useful feature in large castings. They also make machining easier and absorb shrinkage stresses as the casting solidifies. The specimen has been coated with a thin layer of iron oxide in order to reveal the various constituents of the cast iron in different colours. The graphite flakes are deep blue, set in an orange background of pearlite. The deep orange skeleton at centre is ferrite which contains tiny particles of iron phosphide.
LM, polished section, Nital etch, ferric oxide coating, ×950

7.36 Although flake graphite confers some advantages to grey cast iron, it has the drawback of making the casting brittle and susceptible to shock forces. The shape of the graphite can be modified by alloy composition and heat-treatment. The addition of 0.04–0.06 per cent magnesium causes the graphite to form as balls. This is known as 'spheroidal cast iron' and is more ductile than grey cast iron because it lacks the sharp flakes. This specimen has also been coated with iron oxide, colouring the graphite rosettes blue against the magenta of the ferrite. Sweeping across the far right of the picture is a different form of pearlite which has a grainy appearance. The platelets have here been replaced by small spheres of cementite to form spheroidised pearlite.
LM, Nomarski DIC, polished section, Nital etch, ferric oxide coating, ×950

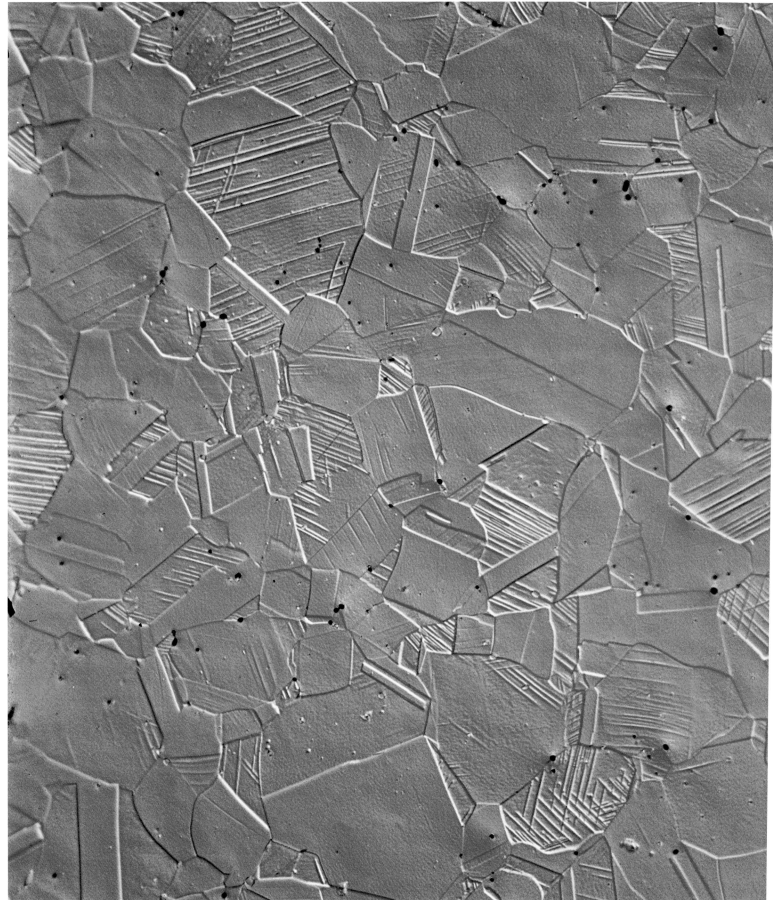

7.38

7.39

Seven representatives of the vast range of non-ferrous metals and alloys have been chosen. The range of alloys available to today's designers has extended the limits of performance of most machines.

7.37 Despite the fact that it is 75 per cent iron, stainless steel is not considered to be a ferrous alloy! Steel changes from ferrite to austenite as it passes upwards through 760 degrees centigrade and changes back as it is cooled. But the addition of 18 per cent chromium and 8 per cent nickel makes austenite stable at room temperature and forms 'austenitic 18/8 stainless steel'. This is the household utensil material which does not rust in use. Stainless steel cannot be heat-treated like ferritic steel and hardening can only be accomplished by cold deformation or 'work hardening'. This induces twinning in the polyhedral grain structure.
LM, Nomarski DIC, polished and etched section, ×280

7.38 The closely spaced twinning in the stainless steel is quite different to that displayed in the pure copper shown here. The specimen has been etched, and photographed in polarised light. The orientation of the crystal lattice within each grain has determined its colour.
LM, polarised light, polished section, ammonium persulphate etch, ×300

7.39 The addition of 11.8 per cent aluminium to copper produces an aluminium bronze, a material which has strength, good working properties, resistance to corrosion, and retains some of the colour of copper to produce a rich golden alloy. The structure shown in the micrograph consists of a fine pearlitic eutectoid which has coloured green, orange and purple in the Nomarski illumination. A grain boundary runs vertically to the right of centre and is delineated by the black globules of impurity oxides.
LM, Nomarski DIC, polished section, ferric chloride etch, ×75

7.40 The aerospace industry has consistently demanded higher performances from the materials it uses, resulting in many new alloys being developed. The fighter aircraft of today use large quantities of titanium alloy because of its high strength at low weight. The Lockheed SR 71 Blackbird has a titanium alloy skin because it is the best material for withstanding the heat developed at the aircraft's maximum speed of 2200 miles per hour. Titanium alloys are also used extensively in the biomedical field, where they have proved to have good biocompatibility. The sample photographed here in polarised, reflected light reveals twinned grains in the areas close to the material's surface at the top of the picture. These result from the alloy being rolled into shape while it was cold; in other words, they are mechanically induced twins.

LM, polarised light, polished and etched section, ×700

7.41 In some materials, the atomic forces which determine the shape of crystals are so strong that they disrupt smooth surfaces during the transition from liquid to solid. When this happens, the kind of faceted structure shown in this specimen of pure aluminium can develop. Surface tension forces normally cause the surface between a liquid and a solid to remain flat and smooth. Here the atomic forces have caused the solidifying atoms to build onto faceted planes because they are more favourable for distributing the energy emitted on solidification.

SEM, magnification unknown

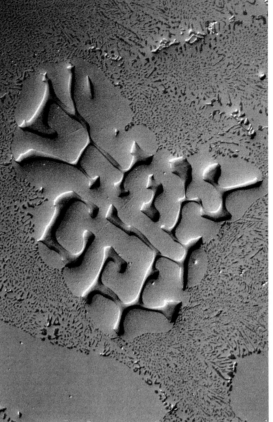

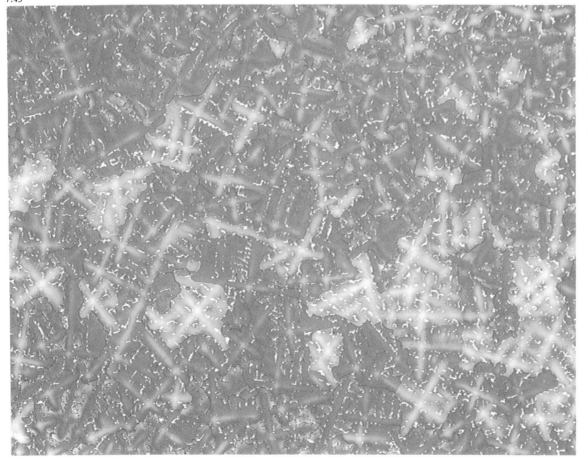

7.42 The addition of silicon as an alloying element improves aluminium's strength and ease of casting. In casting, the molten alloy is poured into a fixed mold so that a complex shape can be formed. The shrinkage on cooling has to be controlled by the selection of the best alloy mix. In this specimen a 'Chinese script' dendritic structure has formed, surrounded by pure aluminium. Outside this region is a fine, mottled eutectic consisting of 11 per cent silicon and 89 per cent aluminium.
LM, Nomarski DIC, polished section, ×1400

7.43 Very demanding environments exist in the centres of jet engines. The turbine blades are moving at nearly the speed of sound in the presence of corrosive, burning gases. Alloys for this type of application must be carefully chosen if the required reliability is to be achieved. Failure can be costly in both human and economic terms. Although they are expensive, combining the metals cobalt and chromium provides the basis for many 'superalloys'. Carbon, tungsten, molybdenum, titanium and tantalum are also commonly added. These elements form hard carbide particles which lock the microstructure together, preventing atomic planes from slipping over each other and conferring good high-temperature strength. The alloy designer is now confronted with a material which is almost impossible to machine, so he must also select his alloy composition for good castability so that the turbine blades can be molded instead of machined. This is why the sample shown consists of a dendritic, cast structure. The white carbides can be seen residing in the interdendritic spaces.
LM, polished section, Beraha colour etch, ×300

CHAPTER 8
INDUSTRIAL WORLD

MICROSCOPES provide a bridge between the world of scientific research and the world of industry. In research, microscopes are employed to explore and develop new or improved materials and processes. Industry takes these developments and attempts to exploit their useful properties to produce real commercial products which perform better, or cost less, than those currently available. In industry, microscopes are frequently used to confirm the quality of a product; and they play a vital role in the analysis of any failures or weaknesses. When an aircraft crashes, microscopy will be an important tool in the investigator's armoury.

A new material may have properties which are so novel that it forms the basis of a whole new technology. The best-known example is the development in the late 1950s of manufactured silicon crystals. The ability to pack thousands of individual transistors into 6-millimetre squares of silicon made possible the modern microelectronics revolution. Video games and 'smart' household appliances, pocket calculators and supercomputers all depend on the ubiquitous silicon microchip. Today, the limits of silicon-based technology have been reached and the learned journals are full of monographs about the new wonder material, gallium arsenide.

Microscopy and electronics enjoy a symbiotic relationship. Microscopes have made possible not only the development, but also the manufacture of microchips. In turn, the microchip industry has provided a spur to the development of new instruments such as the acoustic microscope, which uses sound instead of light or electrons as the imaging radiation. The acoustic microscope is beginning to play a routine role in microchip quality control and failure analysis, because of its ability to see through the chip's opaque surface to the layers underneath.

The quest of microchip manufacturers has always been to reduce the size of their devices. Throughout the 1960s, chip manufacture involved the photoreduction of the patterns of pathways and junctions that are laid down onto the silicon wafers, and this was accomplished with light microscope optics. In the 1970s, however, the patterns became so fine that they could only be made with electron microscope technology. The number of transistors on an individual microchip rose from a few thousand to several million. The electron-beam lithography equipment used to make these modern chips is indistinguishable from an electron microscope to all but the specialist.

It is not only in high-technology fields such as microelectronics that microscopes are in constant use. A great deal of modern industry is based on processes for joining materials together to make components or finished products. Depending on the nature of the materials and the type of join required, different techniques of soldering, brazing, or welding may be appropriate. Cars contain thousands of welded joints, and so do the appliances in a modern kitchen. Checking the quality of such products requires that regular samples of these welds are subjected to microscopic examination. When there are production problems, the numbers sampled are dramatically increased. Quality control of this kind is not always carried out in a leisurely academic manner; when a production line is halted, answers are needed quickly.

The analysis of failure forms a large part of the daily workload of industrial microscopy. The failure of components invariably costs money, whether as a result of repairs or because of warranty claims from irate customers, and in some cases it can cost lives. The extent of the examination depends on the consequences of the failure, but a microscope is almost always used at some stage. The reason is the nature of the atomic features of solids described in the previous chapter. The microscopic marks on a failed component are characteristic of the mode of failure and can be interpreted by the skilled investigator.

A major cause of component failure is corrosion. It has been estimated that in the United States corrosion costs industry 5 per cent of the country's gross national product. The most common method of fighting corrosion is to coat a product with a corrosion-resistant material, such as paint or enamel. Microscopy is closely linked to coating technology. For many years it was the only means of examining and measuring the quality and thickness of a coating, and although other techniques are now available, it still performs a key role.

Amongst the many and varied developments in materials science since the 1960s, one has had a special impact on the quality of life for some people. Bioengineering is the development of materials which are biocompatible and which can therefore be used to replace defective parts of our bodies without being attacked and rejected by the immune system. The replacement hip joint is the most common example. Such *prosthetic* devices need to meet a wide range of criteria. They must not only be biocompatible, they must also have the correct lubricating properties and sufficient fatigue strength to last for years.

8.1 This false colour scanning electron micrograph shows a tiny portion of a 256-kilobyte dynamic random access memory, or DRAM, microchip. The different colours clearly distinguish the layers of pathways that cover the surface of the chip. The movement of electric currents along these pathways forms the basis of the chip's operation. The three light blue pathways on this DRAM are each 3 micrometres wide and have been produced by photolithographic techniques. The latest DRAMs have pathways as narrow as 0.8 micrometres, produced by electron-beam lithography. The half-sunken pad on each of the light blue pathways is an individual transistor memory cell. A memory cell consists of a transistor plus a number of other electronic components. The sheer concentration of information on chips like this, which might typically form part of a computer's memory bank, can be grasped from the quantity of its transistors. A microchip of 256 kilobytes has 256 000 bytes; each byte consists of 8 bits, and each bit – each single unit of binary information, 1 or 0 – requires 1–2 transistors. The whole DRAM thus contains between two and four million separate transistors. A modern mainframe computer might contain 400–1000 DRAMs, together with other types of memory microchips. DRAMs are so-called because they do not store information on a permanent basis; on the contrary, their memory cells need to be refreshed every 2 milliseconds, or two-thousandths of a second. The advantage of this is that fewer transistors are needed to construct a memory microchip. They are also inexpensive, and are therefore used in large numbers in both mainframe and personal computers.
SEM, false colour, ×11 300

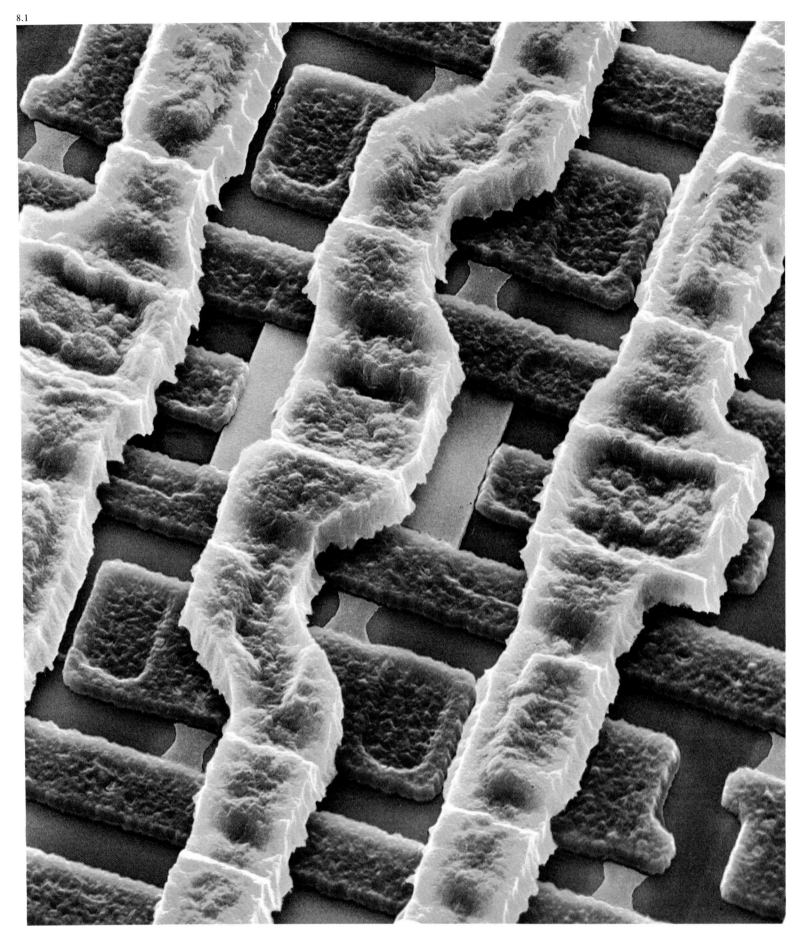

MATERIAL JOINING

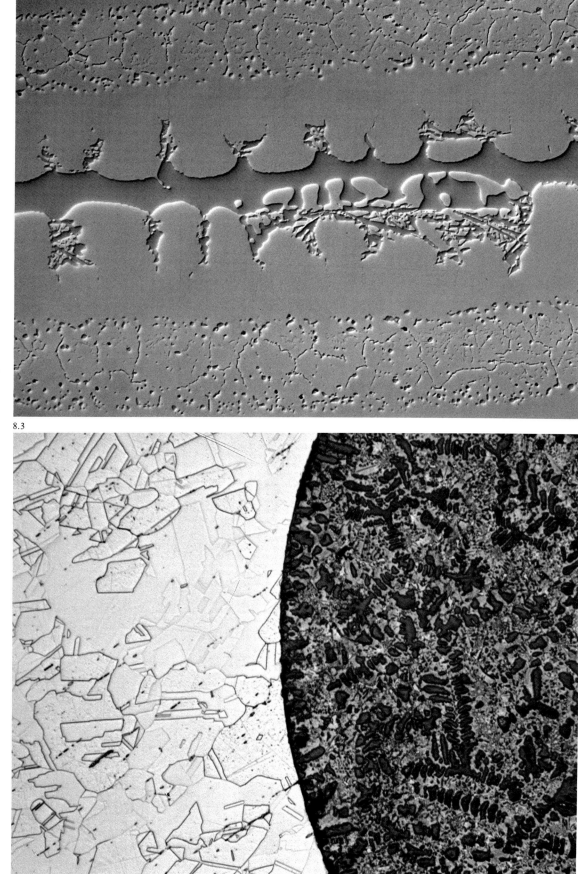

Engineers frequently join materials together in order to achieve otherwise impossible designs, or to maximise the benefits of using special combinations of materials. Metal joining techniques divide into two basic categories: *fusion* processes and *solid phase* processes. Fusion processes include soldering, brazing and arc welding; they always involve melting a metal to form the joint. Solid phase processes involve bringing materials together in such intimate contact that atoms can migrate across the gap between them and form atomic bonds. Examples of solid phase joints are friction welds and explosive welds. Resistance welds are a mixture of both solid phase and fusion techniques, made by clamping two sheets of metal between copper electrodes and passing high electrical currents between the electrodes. This process, also known as 'spot-welding', is commonly used to fabricate the body shells of motor vehicles.

8.2 Some combinations of materials or material thicknesses are difficult to weld. When welding is inappropriate but a strong joint is required, as in this aircraft part, a *braze* is often employed. Brazing alloys usually melt at around 600 degrees Centigrade and are formulated from a variety of silver, copper, zinc and gold alloys. The join line runs horizontally across the centre of the picture; the rugged features on either side of the join are dendrites projecting into the formerly molten zone. These dendrites have grown while the interface was cooling; they are the result of new alloys forming to make the joint.
LM, Nomarski DIC, polished section, etched, ×900

8.3 A frequent requirement in the electrical industry is the joining of copper to other metals. With the correct choice of brazing alloy,

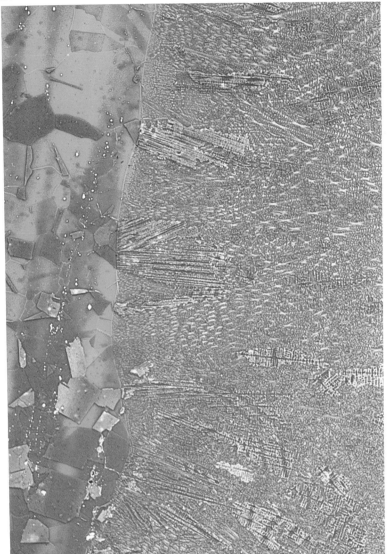

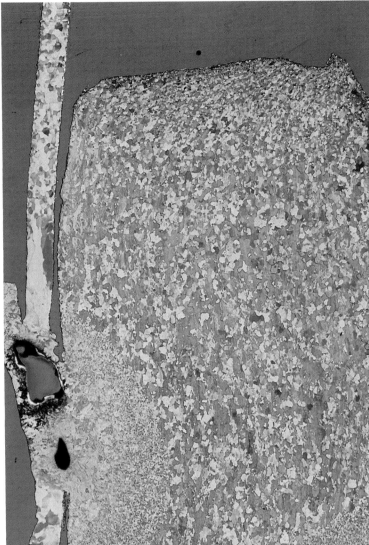

dissimilar metals such as copper and nickel can be joined. This light micrograph shows the interface between the brazing alloy and copper in such a joint. The copper is on the left. The silver-based brazing alloy contains well-formed dendrites, created as it cooled and solidified. The curved interfacial area contains a dark, copper-rich phase, which formed as the silver alloy dissolved copper during the joint-making process.

LM, polished section, ammonia and hydrogen peroxide etch, ×50

8.4 The alloys used to form the joints in soldering and brazing are usually weaker, and less resistant to corrosion, than the parent metals which are being joined. These drawbacks are avoided in fusion welding, where the parent metals are themselves melted so that a molten weld pool forms across the gap between them and then solidifies to form the weld. The melting point of the parent metals must be attained, and this temperature is almost always higher than that used in soldering or brazing. In order to prevent excessive melting, a highly localised heat source is required. A common method is to use electric arcs. And because molten metals usually react with oxygen in the atmosphere, it is necessary to blow 'shield' gases onto the weld during the fusion process. The gases protect the molten metal from oxygen. The micrograph shows the interface between a parent metal (left) and the weld fusion zone. The parent metal is a stainless steel, and has a polygonal grain structure. The weld pool has solidified into a dendritic structure which covers most of the right side of the picture. One advantage of fusion welding is that where two components made of the same metal are being joined, the weld retains the same chemical composition – and therefore the same corrosion resistance – as the parent metal. A disadvantage of fusion welding is that it cannot be used where different parent metals have widely different melting points.

LM, polished section, Beraha etch, ×290

8.5 The microscope is frequently used to examine the quality of welded joints, although this is necessarily a destructive technique which sacrifices the weld. The example shown is a cross-section through a defective resistance weld on an electronic component. The weld is intended to join the thin sheet of iron on the left to the thicker sheet on the right. Incorrect conditions at the interface between the welding electrodes and the thin sheet have caused metal 'splattering', resulting in the two irregularly shaped holes, the largest of which has almost severed the thin sheet. The micrograph also demonstrates other features. The area of fine-grained material radiating from the smaller hole is the heat-affected zone, where the heat from the weld interface has caused the iron to recrystallise. Recrystallisation into a finer grain size is also evident at the top of the thick sheet, indicating that it has been stamped into shape. The distortion of the crystal structure during the stamping has caused it to recrystallise at a later heat-treatment stage. The small projection in the top right part of the thicker sheet is a 'shear lip'. Shear lips are also associated with stamping processes and are responsible for the cut fingers frequently sustained when attempting to service cars or washing machines with unfinished metal edges.

LM, polished section, Klemm etch, ×70

SOLID PHASE JOINING

It is not essential to melt two surfaces together to create a welded joint. In solid phase welding, no melting takes place; the join is achieved by bringing the surfaces together in intimate atomic contact. The natural attraction between atoms creates extremely strong bonds, especially if some heat is applied.

Friction welding is the most common solid phase welding process and is used in a wide variety of applications, notably in the automobile industry. To make a friction weld, two bars of metal or plastic are rotated in contact with each other so as to generate frictional heat. During this part of the operation, one bar is slowly pushed into the other so that clean metal is 'burned' onto the weld surface. The rotation is then stopped very quickly and the bars are forced together to forge the hot, clean interface.

8.6 One advantage of solid phase welding is its ability to join dissimilar metals. This cannot be done by fusion welding if there is a large discrepancy between the melting temperatures of two materials, such as nickel and aluminium. This scanning electron micrograph shows a nickel-aluminium joint which could only have been made by microfriction welding. Small welds such as this have to be made at very high rotational speeds in order to generate sufficient surface heat. This weld would be rotated at 60 000–100 000 revolutions per minute before being stopped in microseconds and forged together. The three protrusions at the top are the result of the forging force squeezing aluminium between the jaws of the holding chuck. These protrusions and the circular 'welding flash' would normally be machined off before the component was used in service.
SEM, ×30

8.7 It is usual when examining the microstructural quality of a friction

weld to cut and polish a cross-section. This example shows a good-quality weld between a pure iron plate (bottom) and a low-alloy steel pin – there are no voids or inclusions in the weld interface. Joining pins and tubes to plates is normally a difficult task, but it is made relatively easy by the use of friction welding. The plate is held stationary while the pin is rotated against it. Any rust or impurities on the surface of the steel plate are forced out into the weld flash, leaving the interface clean and sound. The colours of the grains have been produced by the chemical etching, which has stained them.
LM, polished section, Klemm etch, ×150

8.8 The light microscope is commonly used to study defects in welds. In this friction welded specimen the rod at the top and the plate at the bottom are separated by a horizontal tear. A failure of this kind is often the result of incorrect machine settings. If the forging pressure is applied before the rotation has stopped, for example, the freshly made weld will be torn apart. In this picture the disturbance to the grain structure from the frictional forces is shown. The vertical grain alignment of the rod at the top of the picture has been changed to a fine, brown structure which has spread out horizontally. The grain structure of the plate was already aligned horizontally and so the change is evident only by the colour change from blue to brown. The colours are the result of the chemical etch.
LM, polished specimen, ammonium molybdate etch, ×290

8.9 Explosive welding is another type of solid phase welding and one of the most spectacular. It is carried out between two plates of metal, one of which is held stationary while the other is literally fired into it by an explosive charge. The force of impact is so great that the metals momentarily behave like liquids, creating waves at the welded interface. Such waves, complete with collapsed crests, are seen in this micrograph of steel plates coated with alternate layers of copper and nickel.
LM, ×100

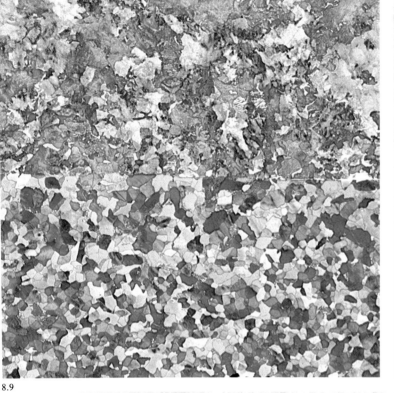

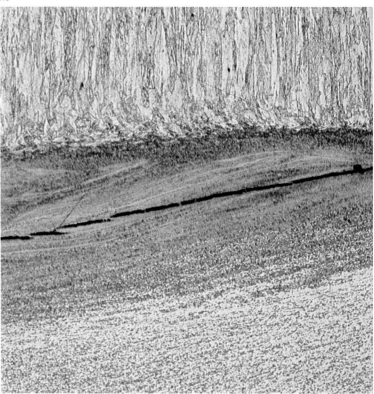

8.9

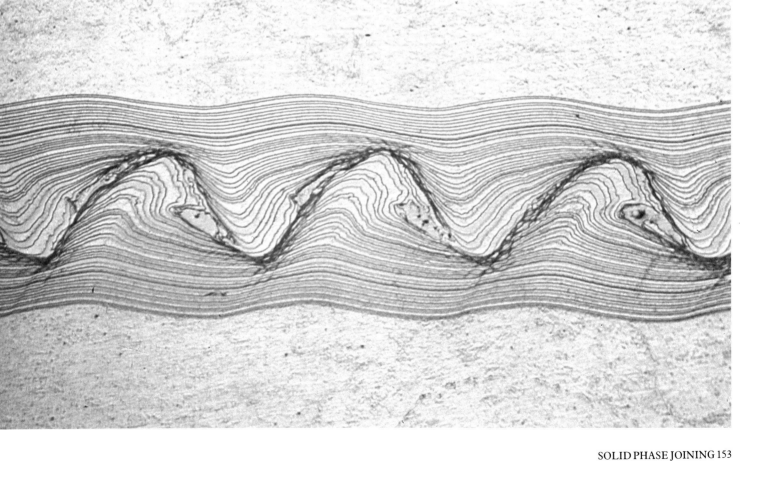

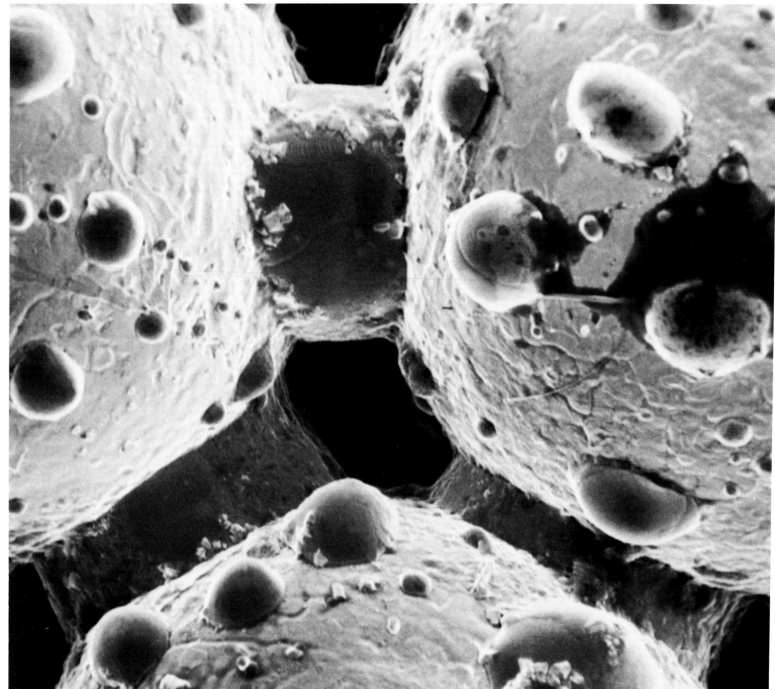

EXTREME DUTY
MATERIALS

Many modern environments and duties subject materials to conditions that are enormously demanding. An example is the interior of a jet engine – an inferno

in which extreme heat, stress and corrosiveness are combined. Yet we demand that the turbine blades in jet engines are manufactured to an extremely high standard, because the consequences of a failure can be catastrophic. Applications like this have led to the creation of a number of 'extreme duty materials'. In other cases, new materials have been found which have unique properties, and this has

encouraged the development of new applications which can exploit their special qualities. An example is boron carbide, which is so hard that it is an equally effective – but cheaper – alternative to diamond abrasives.

8.10 The metal with the highest melting point is tungsten and it is employed for its heat resistance in such demanding applications as the exhaust cones of rocket motors. It is a difficult material to utilise, partly because its

melting temperature of over 3300 degrees Centigrade means that few materials can form a crucible in which to melt it. A variety of methods have been developed to utilise the capabilities of tungsten. This scanning electron micrograph shows small spheres of tungsten bonded together by a copper 'glue'. The copper forms both the bridges between the spheres and the globules adhering to their surfaces.
SEM, ×830

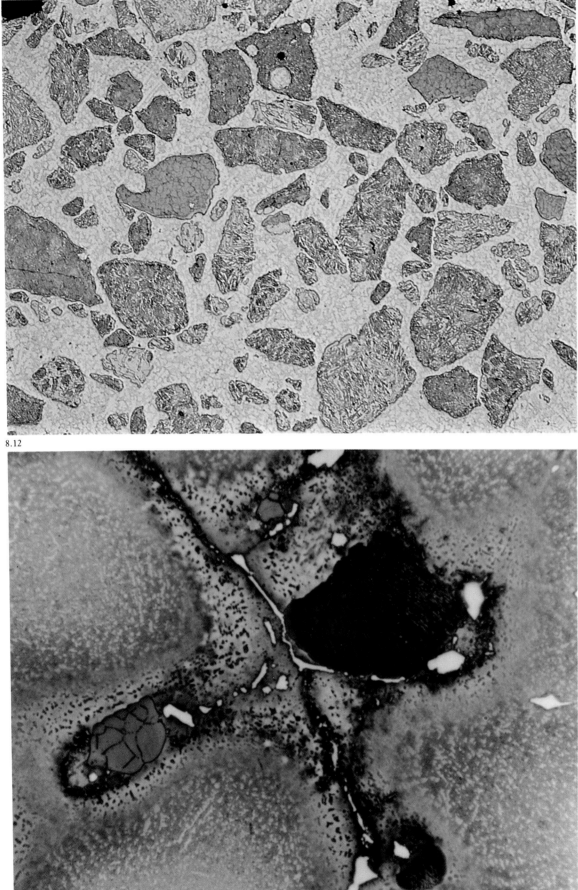

8.11 When tungsten is combined with carbon, a material of extreme hardness is produced. Tungsten carbide is capable of machining hard steels at high speeds and is very important as an industrial cutting tool. But it suffers from the drawback of being brittle. This is overcome by binding tungsten carbide particles with a copper filler. In this micrograph, the tungsten carbide particles are coloured and the copper is almost transparent. The hard, sharp, carbide particles create and maintain the cutting capability, while the copper has a very high heat conductivity and can therefore dissipate the large amounts of heat generated in a tool tip as it cuts.
LM, polished section, ×220

8.12 The turbine blades in modern jet engines are driven by corrosive gases burning at very high temperatures. The blade tips travel at almost the speed of sound and experience massive centrifugal forces. Because of these conditions, they are made from special blends of alloying elements. The metals chromium, cobalt, nickel and aluminium usually feature in the mix, together with the carbide-forming elements tungsten, molybdenum, titanium and tantalum. This specimen from a turbine blade has been etched to show the alloy's constituents. The medium and dark blue regions are a nickel–aluminium–titanium phase (the same as in the Kirkuchi pattern in Figure 7.4), while the white granules are the mixed carbides. The large grey areas are the arms of dendrites formed when the turbine blade was cast. Microscopy plays a role in both the design of alloys for turbine blades and the close scrutiny the blades undergo during manufacture.
LM, polished section, Weck's etch, ×1000

CERAMICS

Ceramics are very stable materials, which is one reason why they have often survived from antiquity. The majority are oxides of metals and the range of properties they possess is huge. Almost all are electrical insulators, have high-temperature strength and are resistant to corrosion. Ceramic products range from kitchenware to sophisticated technical ceramics used on microchips or in the tiles which cover the underside of the space shuttle. The most serious drawback of ceramics is their brittleness, but even this is being addressed by the addition of toughening chemicals.

8.13 In metal purification processes, chemicals called fluxes draw the impurities from an ore to make the pure metal and a ceramic by-product called slag. The development of this technology in ancient times helped to advance the art of metallurgy. The slag in this micrograph comes from a copper-making process that took place in Mitterberg, Austria, in 1600 BC. The section contains three circular gas bubbles in a matrix of fayalith, an iron silicate mineral. This is partially *devitrified* and consists of black granules of crystalline fayalith contained within the almost transparent 'glassy' or vitreous form.
LM, polished section, ×400

8.14 Silicon carbide has a variety of useful properties, including electrical conductivity. Although it is classed as an engineering ceramic, the geologist knows it as the mineral carborundum, the metallurgist uses it for furnace linings and electrical heating elements, and the jeweller uses it to polish gemstones. The section shown is from a *refractory brick* of the type used to line the walls of metal-smelting furnaces. The specimen has been electrolytically etched in oxalic acid to reveal its crystalline structure.
LM, magnification unknown

8.13

8.14

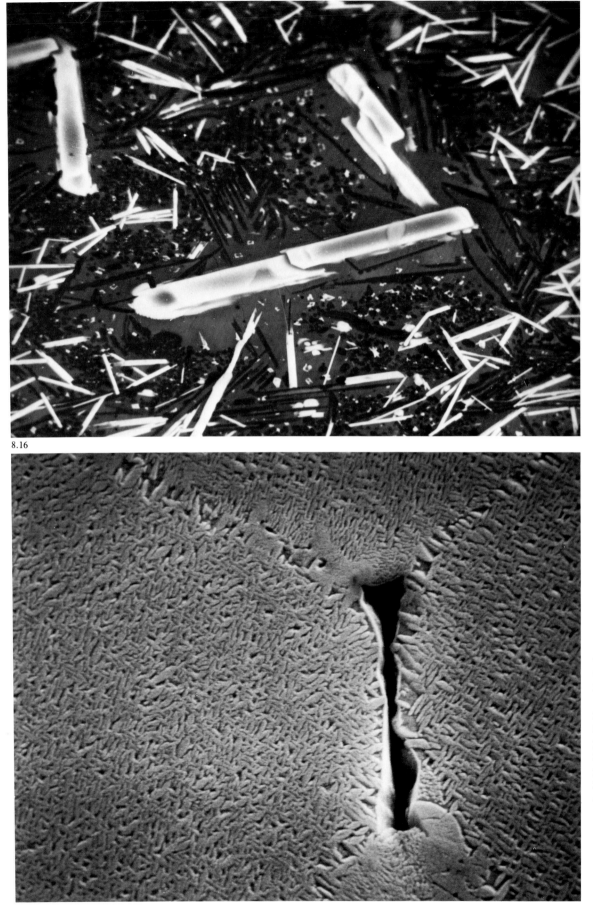

8.15 The silicon carbide in the previous picture and the silicon nitride illustrated here are both members of a group of ceramics that have become known as 'engineering ceramics', because of their use in applications such as novel types of car engine. This silicon nitride has been subjected during fabrication to a strengthening process called 'hot-pressing', in which the last remains of porosity are literally squeezed from the ceramic at high temperature. The unusual appearance of the scanning electron micrograph is due to 'back-scattered' electrons – instead of the usual secondary electrons – being used to produce the image. Back-scattered electrons are primary electrons which are reflected by the specimen – instead of being absorbed by it. They are collected by a special detector and can be used to identify particular constituents in materials. The white needles in the picture are yttrium–silicon oxides, the black needles are the mineral cristobalite, and the dark grey areas are non-crystalline, glassy phases.
SEM, back-scattered mode, polished section, ×350

8.16 One of the most exciting ceramics to emerge in the 1980s is partially stabilised zirconia, which is based on zirconium oxides. It was discovered to have the ability to change phase structure from tetragonal to monoclinic, depending on the temperature and stress to which it is subjected. The change involves an expansion of the ceramic, which fills cracks and prevents them from moving. This makes zirconia both stronger and less brittle and has earned it the nickname of 'ceramic steel'. The micrograph shows an intergranular pore with grain boundaries radiating from its top and base. Most of the picture consists of the lozenge-shaped zirconia grains.
SEM, polished section, phosphoric acid etch, ×25 000

BIOENGINEERING

Spare-part surgery is one of the most revolutionary advances in modern medicine. It is made possible by 'bioengineered' materials which can perform biological functions. Replacing tissue, from heart valves to hip joints, with 'foreign' materials has raised many problems. The most obvious is that the implanted material must be biocompatible to prevent rejection. It must also be able to perform its designated role over long time periods.

8.17 When metal components are needed for bioengineering, titanium is one of the first choices. It has high biocompatibility combined with good strength and low weight. This specimen is a titanium alloy of the kind used in hip replacements. The titanium is cast or forged into the ball-shaped part of the joint which fits into the top of the femur. The 'basket weave' structure exhibited by the alloy is an arrangement of alpha and beta titanium phases generated by rapid quenching during heat-treatment.
LM, polarised light, polished section, ×460

8.18 The ball-shaped part of a hip joint fits into a 'cup' in the pelvis, and this alumina bone implant is used to form the cup in hip replacements. It is a 99.9 per cent pure ceramic and its minimal porosity, illustrated by the very small intergranular pores, gives it good mechanical performance.
SEM, polished section, thermal etch, ×21 000

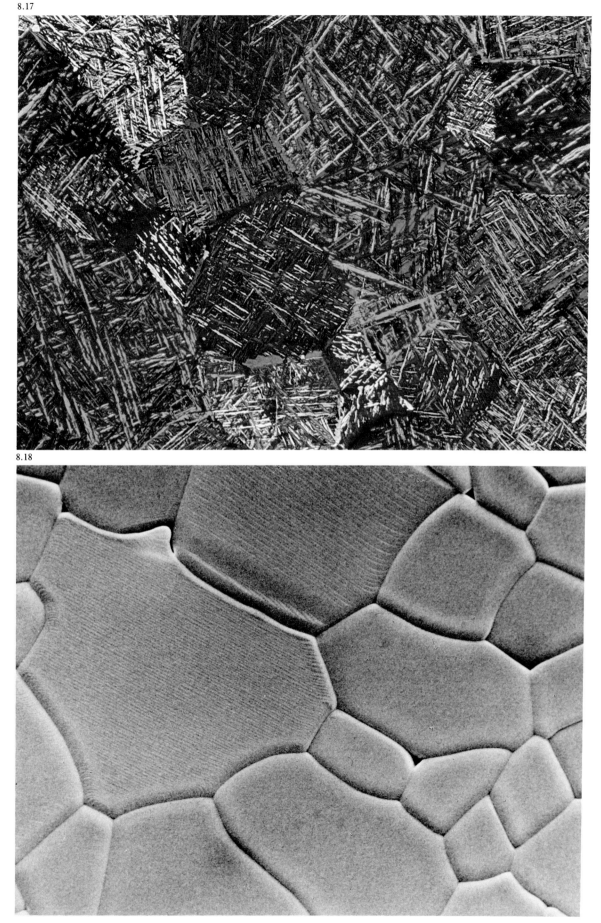

8.17

8.18

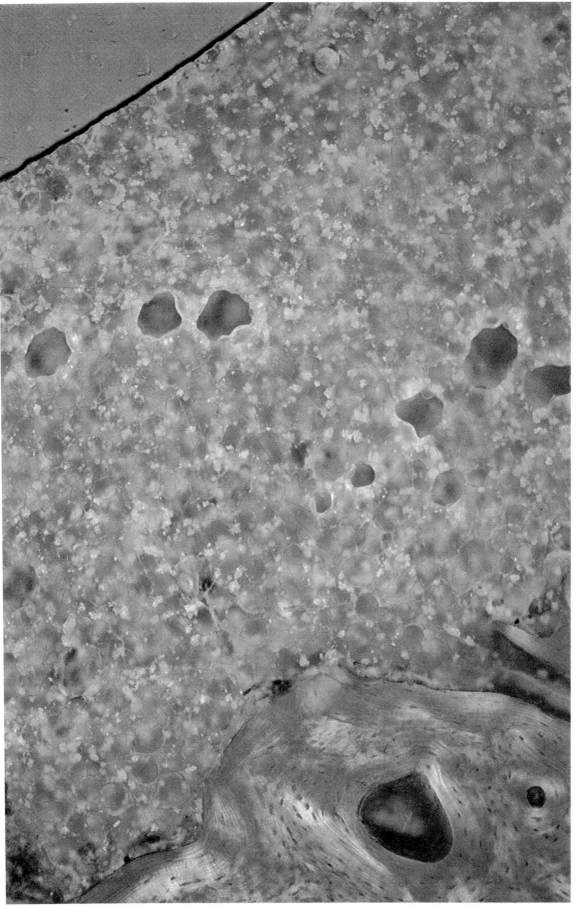

8.19 A major problem in hip replacement operations is attaching the ball-shaped prosthesis to the extremity of the existing bone. The ball usually has a shank which is cemented into the hollow core of the femur. This light micrograph shows the interface between living bone, at bottom right, and cement. The pink triangle at top left is the medium in which the specimen was mounted.
LM, polarised light, ×580

ELECTRONICS

The discovery of the transistor in 1947 signalled the start of a technological revolution. Two transistors connected together in so-called 'flip-flop' mode can electronically represent the binary states 'yes–no', 'on–off', '0 or 1'. This gives groups of transistor pairs the ability to follow logical sequences, perform calculations and store information. In 1957, integrated circuits – or microchips – were invented. They are based upon the special properties of materials called semiconductors,

notably silicon. Microchips are typically 6 millimetres square and less than 0.5 millimetres thick, but they are complete electronic units which can contain millions of transistors. This size and 'device density' gives microchips their versatility and power.

Microchip manufacture begins with the slicing of thin, circular wafers, 75 millimetres in diameter, from long bars of single-crystal silicon. Each wafer forms a substrate onto which about 100 separate microchips are made by a succession of coating and etching processes. First, a layer of 'doped' silicon is deposited in perfect crystallographic alignment with the substrate. The dopants –boron and gallium are examples – either substitute for some of the silicon

atoms in the unit cells or take up residence in the spaces between them. They give the deposited layer the degree of electrical conductivity required by the microchip.

The etching and coating stages are carried out through masks, like stencils in miniature. Where the mask is open, chemicals can etch the exposed area or additional semiconductor material can be deposited. The complex criss-cross network of a microchip's circuitry is built up by using a number of different masks. Transistors and other electronic devices are built vertically and consist of three or more layers.

8.20 This micrograph shows the junction of three memory microchips

and a test module (right) on a manufactured wafer. Electronic probes descend onto the test module's pads as part of the quality control process. Once testing is finished, the individual microchips are cut from the wafer by running a diamond scribing device along the bands between each chip. LM, ×135

8.21 The extraordinary complexity of many microchips is illustrated by this microprocessor. Microprocessors are the central processing units of a computer, packed into a single microchip. As the picture shows, they consist of a great variety of devices. Prominent are the square pads around the edges of the microchip. Some are test pads; others will have connecting wires welded to them to provide the microchip's links with the outside world. The microprocessor shown is a STL80 manufactured by STC. LM, magnification unknown

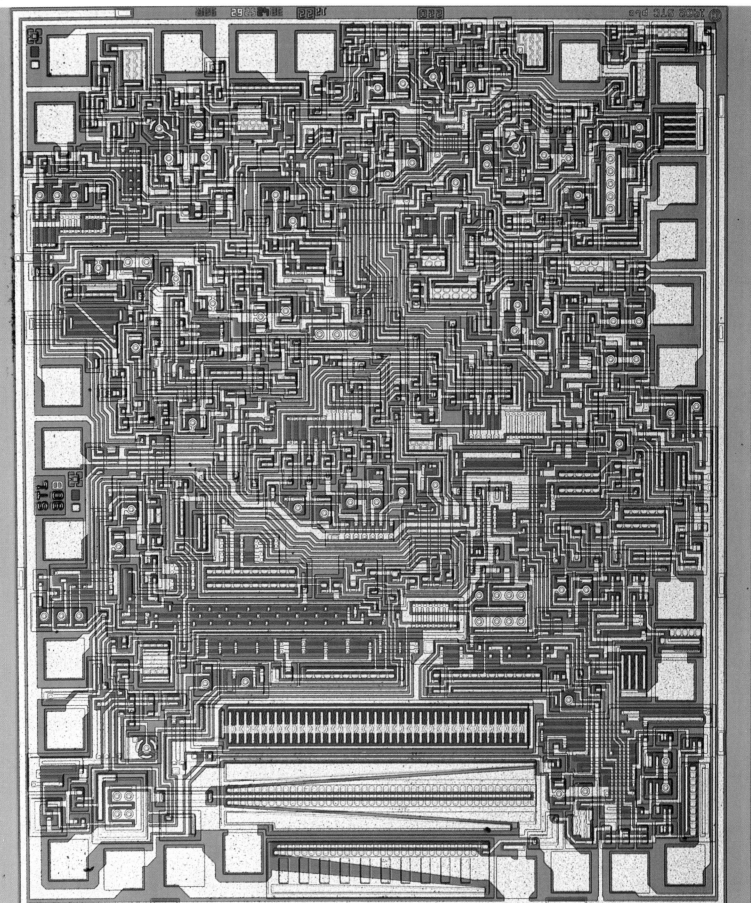

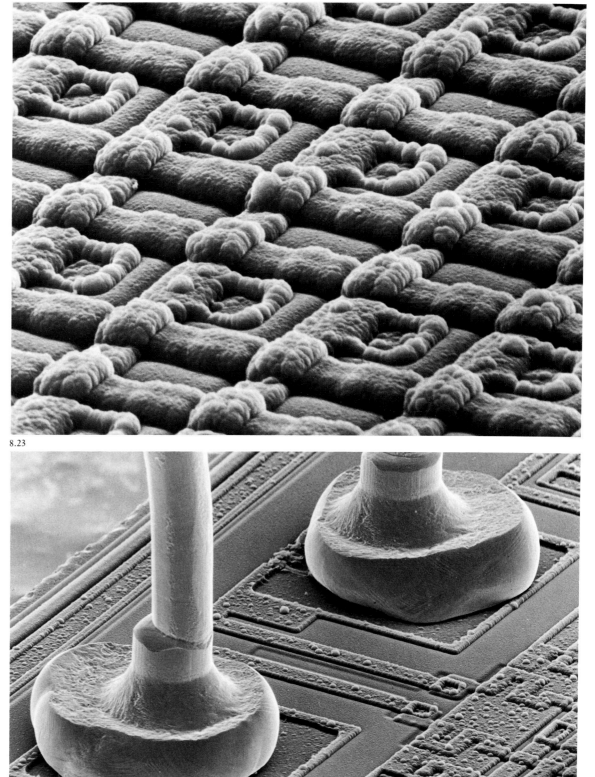

8.22 This scanning electron micrograph shows the memory cells of an erasable programmable read-only memory microchip – an EPROM. The memory devices are the 'p'-shaped features which are repeated across the picture. They provide 16 384 bits of memory per microchip and can be switched electrically so that the microchip remembers and stores the information programmed into it. By shining ultraviolet light through a quartz window in the microchip's outer packaging, the memory can be erased and made available for reprogramming. A typical application is the specialist dictionary that can be attached to some word processors. Many word processors incorporate a standard dictionary to check word spellings, but EPROMs can be added which contain technical words; such microchips work much faster than tapes or discs which previously served this function.
SEM, ×1280

8.23 This is a lower magnification view of another part of the EPROM microchip seen in the previous picture. It shows two connecting wires welded to terminal pads on the edge of the microchip. The wires are attached by ultrasonic welding, a process in which they are rapidly rubbed and pressed onto the pads. Between the pads, two narrow tracks lead towards the microchip's circuitry at bottom right. The small squares at the right-hand end of these tracks are the transistors which control the inputs from and outputs to the two connecting wires. Scanning electron microscopes are used extensively to ensure the quality of microchip welded joints and the accuracy of their placement. An error of just 0.06 millimetres would have caused these connecting wires to be welded onto the main circuitry of the EPROM instead of the terminal pads.
SEM, ×275

8.24 This micrograph shows the whole of the EPROM microchip that is featured in detail on the page opposite. It is seen here sitting in a well in its outer packaging. The connecting wires link the microchip's terminal pads to conductors on the packaging, where they are also welded. As more complex microchip designs have evolved, the number of connecting wires has increased. But not all the termination pads have wires welded to them. The empty pads are used for checking the microchip electrically before it is cut from the wafer. Many microchips are rejected at the final stages of manufacture because of microscopic defects on their surfaces or at junctions. This is why microchips are made in 'clean rooms' with filtered air and specially dressed operators. A simple thing like dandruff can play havoc with a batch of microchips if it escapes from beneath an operator's protective cap.
SEM, ×12

8.25 One of the factors limiting how small a microchip can be made is the heat generated by the passage of electrical current through its circuitry. In order to overcome this, silicon is frequently bonded directly to the microchip's outer packaging so that the heat can dissipate more readily. The quality of the bonding is important in such cases and a number of techniques have been tried to test it non-destructively. One of the most promising is the scanning acoustic microscope or SAM. The SAM creates an image by using sound waves, which have the valuable property of being able to penetrate beneath visually opaque surfaces. In this micrograph, the sound waves are focused beneath the surface of the microchip, seen as the red square, to show the good-quality central bond, coloured blue. The picture also shows well-bonded (blue) and poorly-bonded (red and yellow) conductors on the packaging surrounding the microchip.
Scanning acoustic micrograph, false colour, ×10

8.26

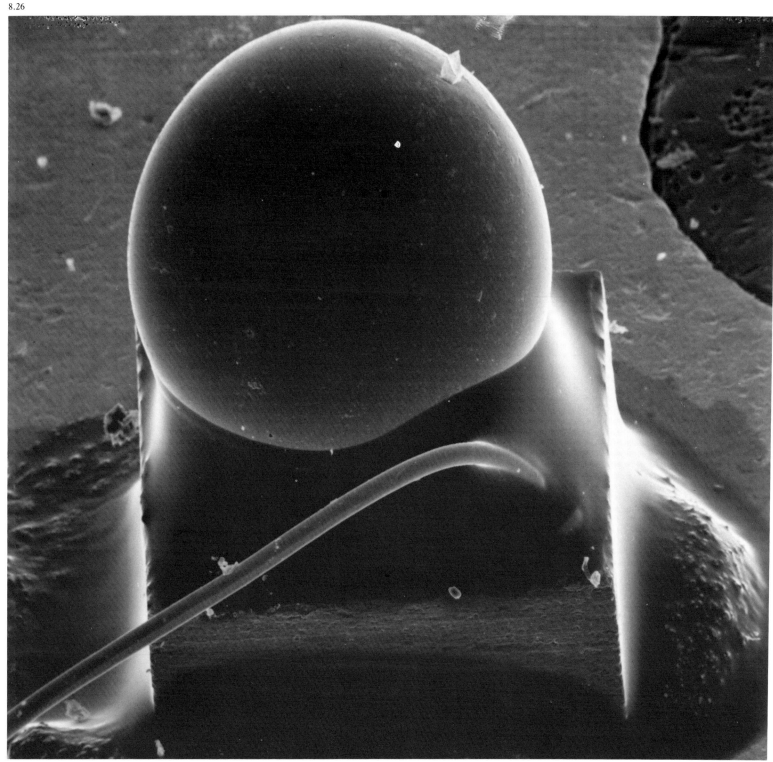

8.26 A new electronics material emerged in the early 1980s to challenge silicon's supremacy – gallium arsenide. Coated onto sapphire (aluminium oxide) substrates in so-called 'thin film' technology, it has the ability to make microchip devices operate 5 times faster than its silicon-based counterparts. This is a major advantage in supercomputers, where the speed of the devices influences the speed at which calculations can be performed. Another property of gallium arsenide is that in a particular combination with gallium-aluminium arsenide it can be used to make very small, fast-acting devices which emit light. An example of such a light-emitting device is shown in this scanning electron micrograph. The light output is of high spectral purity and it can be switched on and off 25 million times per second. Both these qualities make it ideal for digitally encoding phone conversations or other telecommunications signals which are to be transmitted down optical fibres instead of conventional copper cables. The light-generating junction within the device is situated beneath the glass bubble. The latter acts as a lens to focus the light to a fine spot, which has led to the device being termed a 'sweet spot'. The wire at the base of the bubble is the connecting wire through which the electrical input arrives to excite the light-generating junction. SEM, ×105

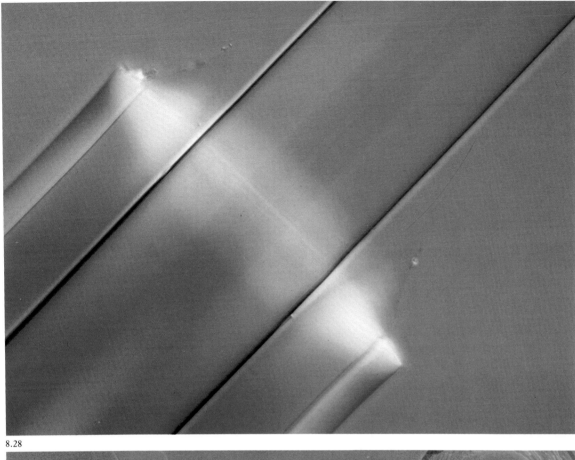

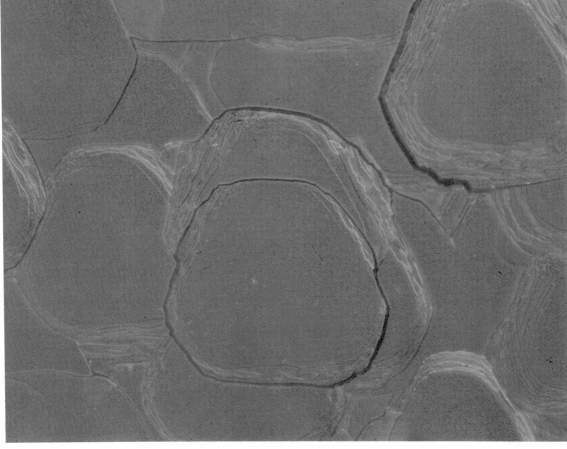

8.27 The core of an optical fibre is made of glass with a refractive index different to that of the glass which surrounds it. The effect of this is that light travels down the core of the fibre, minimising the loss of light through its surface. In this light micrograph, the core of the fibre shows a slightly deeper pink, or blue in the bottom left corner. The additional external tube at left is the fibre's protective cladding; it plays no part in the optical transmission but protects the delicate glass fibre from mechanical damage.
LM, interference contrast, ×385

8.28 Analogous to the light-*emitting* properties of gallium arsenide devices are the light-*absorbing* properties of solar cells. Solar cells absorb radiant energy from the sun and convert it into electrical energy. Most solar cells are made of single-crystal silicon, although thin coatings of silicon, gallium arsenide and cadmium sulphide are also used. This light micrograph shows the surface of a silicon crystal solar cell. The substrate consists of one type of silicon (called n-type) and onto it another type of silicon (p-type) has been deposited and formed in stepped platelets. High cost initially limited the use of arrays of solar cells to esoteric applications such as powering space satellites. As their price fell, they became more widely used – to heat swimming pools, for instance, and even to power wristwatches. Larger arrays are used in remote areas for standby power applications, and there are many prototype arrays around the world investigating their possible use to provide electricity for domestic purposes.
LM, Nomarski DIC , ×900

CORROSION

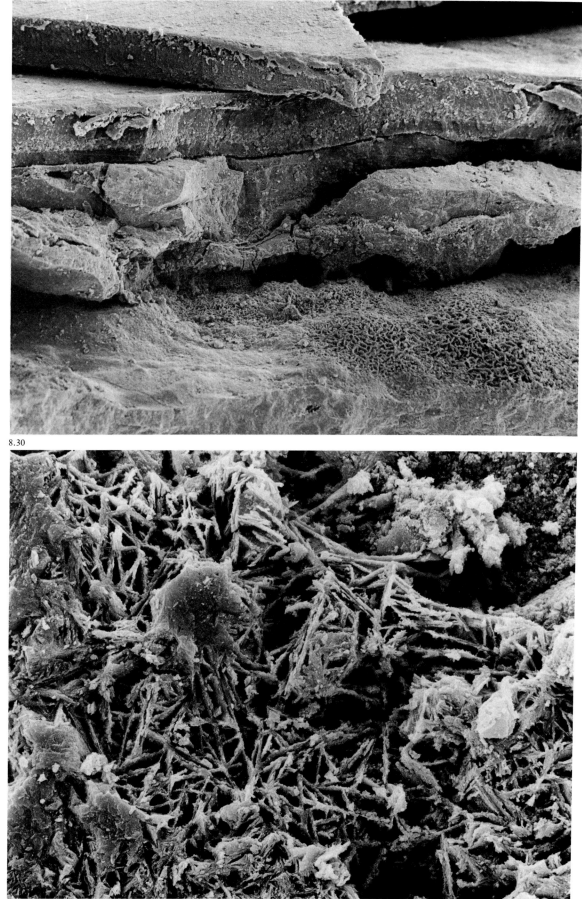

Corrosion costs billions of pounds each year. Although it takes many forms, it is usually the manifestation of an undesirable chemical reaction. In the case of common rust, it is the oxidation of steel; with old buildings, it is the attack on the stonework by pollutants such as sulphur dioxide; in chemical plants, it can be caused both by external, atmospheric sources and from the inside, as a result of the substances that are processed or stored within the tank.

Rust is the most common form of corrosion because steel is the most widely used metal. Rust forms when water and oxygen, the two corrosive agents, combine to react with iron. The oxide forms on the surface of the steel and expands until the stress this creates causes the top layer to spall off and expose fresh steel. In some instances it is the expansion itself, rather than the loss of metal, which causes the damage. For example, when an iron or steel bolt in cast concrete begins to rust, huge pressures build up which eventually shatter the concrete.

8.29 This picture shows a piece of bodywork from a 12-year-old motor car. No metal is seen. The triangular fragment at top is a flake of paint from a poor respray; beneath it are the three layers of paint sprayed on by the manufacturer. The lower third of the micrograph consists of rust.
SEM, ×75

8.30 The crystalline nature of the surface of rust is shown in this detail of the specimen described above. Rust grows when moisture catalyses the reaction between iron and atmospheric oxygen. The reaction product is hydrated ferric oxide, and in this case it has grown along preferred planes to produce a crystal structure.
SEM, ×825

8.30

COATING

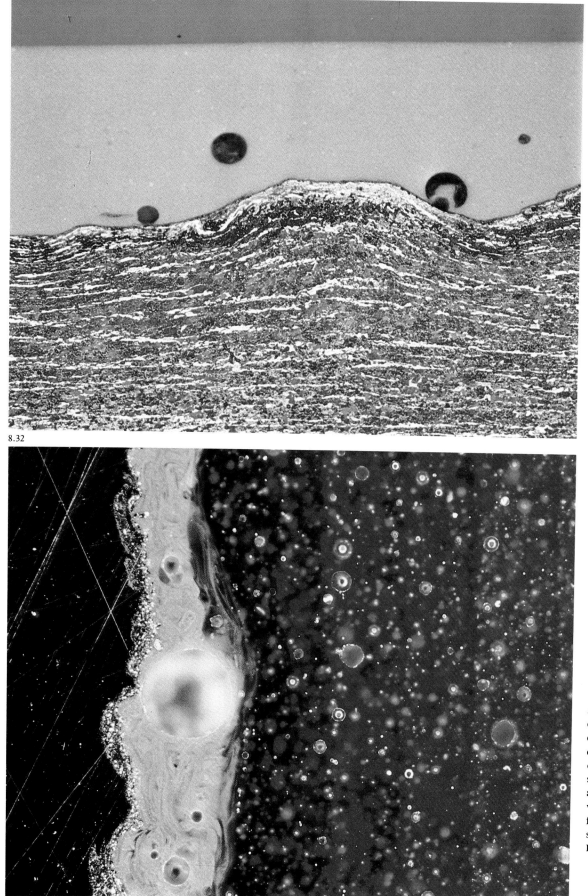

The most common defence against corrosion is to coat a component or product with a thin protective layer of a non-corroding substance. Many sophisticated coatings are available today, but the vast majority of products still rely on the basic technologies of lacquering, painting and enamelling.

8.31 This light micrograph is of a cross-section of lacquered steel. The steel is at bottom, with the colourless lacquer above it. The lacquer, which has been applied by spraying, contains four air bubbles that appear black. The good adhesion between the steel and the lacquer can be seen from the close contouring of the coating to the microscopic undulations in the surface of the steel. This is essential if the coating is to perform its function of keeping moisture away from the steel. LM, polished section, Klemm 1 etch, ×460

8.32 Enamel is one of the most durable and effective of coatings. It consists of a thin layer of glass, which has the hardness necessary to withstand the domestic abrasives frequently used for cleaning. Because glass does not directly adhere to most substances, enamelling usually involves an intermediate layer or 'interlayer' which binds chemically both to the steel, or other material from which the product is made, and to the glass. Once the interlayer has been applied, the glass is sprayed on in a powdered form called *frit*. The baking of the frit fuses the glass to the interlayer. This cross-section of an enamelled steel shows all three components: the steel in black, the cream-coloured interlayer, and the dark blue of the enamel. The circular structures in the interlayer and enamel are due to bubbles. Slightly darker vertical bands divide the enamel into five layers, which correspond to five sprayings of frit. LM, ×450

FAILURE ANALYSIS

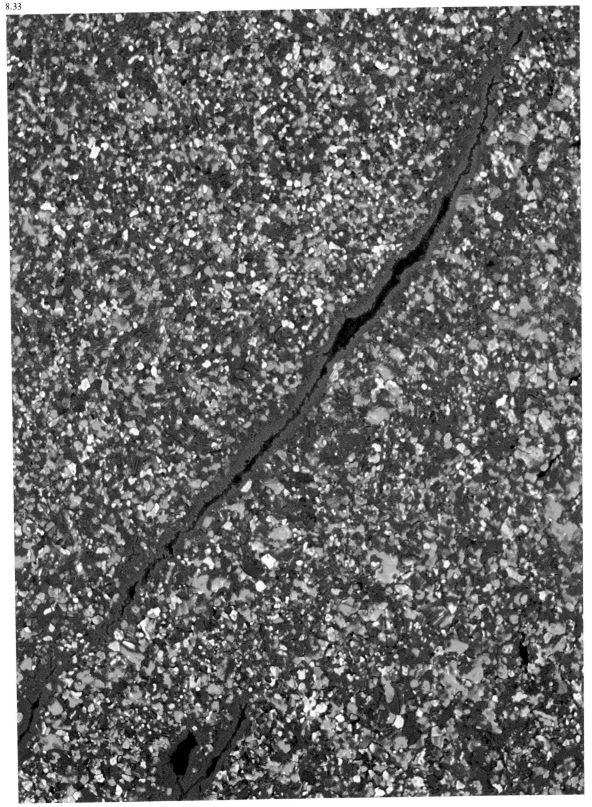

The premature failure of devices and machines is a common occurrence. The effects of failure range from the mild inconvenience of a kitchen utensil breaking to the catastrophic loss of life when a passenger aircraft crashes. The level of post-failure analysis will vary accordingly. The broken handle of the frying pan will rate little more than an angry exclamation and a cursory inspection, but the air crash will be followed by months of intensive detective work.

Visual examination of the fragments produced by a failure plays a vital role in most investigations. The pieces of a crashed aircraft are recovered and painstakingly arranged according to their place in the original whole. The failure is reenacted in reverse and in slow motion. As clues begin to emerge, the original cause of the failure is tracked back to a particular component, which is then examined with all the microscope power available to the modern investigator.

A large part of industrial microscopy is to do with the examination of failed components. As a result, it plays an important role in the quest to give us safer transport, stronger materials, more reliable engines. The next four pages display some of the imagery of 'fractography', the science of fractured surfaces. A vital element in fractography is the fact that most breaking, tearing and rupturing processes leave identifiable and characteristic traces in their wake.

8.33 When the reason for a failure cannot be discovered by examining the surface of a fracture, a cross-section has to be prepared. Because this is a technique which destroys the component concerned, it is usually applied only after all available evidence has been obtained by other methods. This micrograph shows a section from a pressure tube made of a titanium–aluminium–vanadium alloy. The tube, of the kind used in heat exchangers for chemical processing plants, failed during pressure testing. A relatively large crack can be seen running diagonally across the picture, and a smaller crack is visible at bottom left. They were caused by intergranular tearing in the recrystallised microstructure. Although titanium has excellent corrosion resistance, it is susceptible to many common contaminants – including greasy fingers and soaps – during its heat-treatment.
LM, polarised light, polished section, ×580

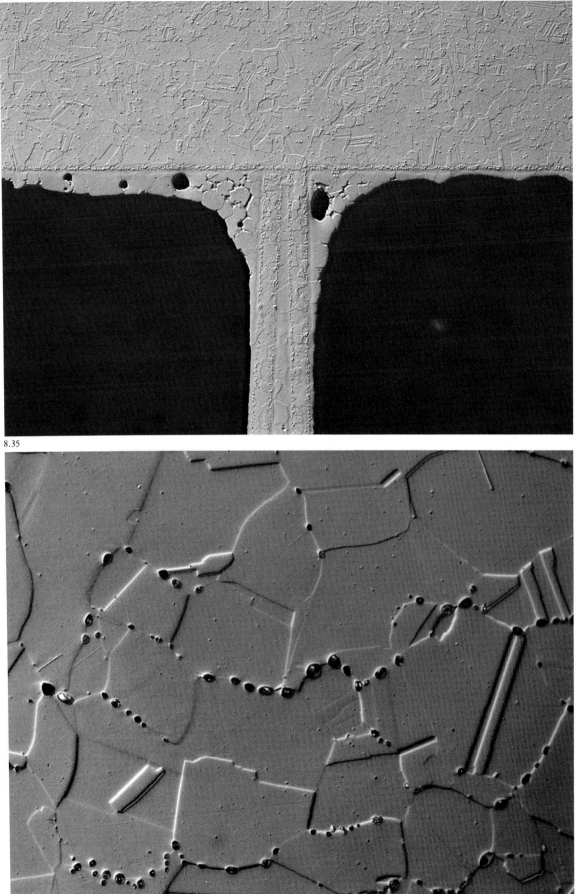

8.34

8.35

8.34 This micrograph reveals serious weakening in a brazed joint in an aircraft component. The brazing makes a 'T' joint between two pieces of nickel alloy, one forming the thick horizontal band across the top and the other the thin vertical strip that divides the lower half of the picture. The brazing alloy forms the two curving 'fillets', slightly darker grey in colour, which border the black background areas. The weakening of the joint is revealed by the small black globules within the fillets. They are indicative of excessive porosity, which could cause the joint – and hence the whole component – to fail.
LM, Nomarski DIC, polished section, etched, ×220

8.35 The fault represented by the black 'beads' in this light micrograph led to a serious explosion at a power station in Australia in the 1960s. The beads are holes in the microstructure of copper, and they are formed if copper containing copper oxides is inadvertently brazed in a furnace using hydrogen as a shield gas. The hydrogen reacts chemically with the copper oxide to produce water molecules and these turn to steam at the brazing temperature and literally blow holes in the microstructure of the copper. The holes are formed along grain boundaries and seriously weaken the copper. It was the use of this type of copper and furnace atmosphere that caused the Australian accident. The electrical fuses, which were supposed to protect the installation, contained brazed connecting tags which had been weakened. These fell off and caused so much arcing on the outside of the fuse that the gases created exploded. Fuses are intended to withstand arcing internally but the failure of the external tag rendered them useless.
LM, Nomarski DIC, polished section, etched, ×220

When a material is undergoing rapid, catastrophic failure, the fracture surfaces that are created are the result of highly organised effects. Even the smashing of a dropped milk bottle produces a mass of evidence for the trained microscopist. A crack moving away from the source of a failure travels quite slowly at first but accelerates rapidly to approximately half the speed of sound for the material involved – usually about 1500 metres per second. As it accelerates the energy driving the crack is dispersed in characteristic ways: first when the crack starts to weave up and down instead of travelling straight, then when it bifurcates to form two cracks. The bifurcation may then be repeated over and over again until the component fragments – hence the dozens of pieces created when the milk bottle falls on the doorstep.

Brittle fracturing, as this process is known, occurs most commonly in ceramics, glasses and hard materials. It can also occur in steels and alloys, but in these materials it is usually preceded by a more leisurely fracturing which is accompanied by deformation of the metal. An example of this phenomenon is when a paper-clip is repeatedly bent until it breaks in two. The bending damages the metal by hardening it – 'work hardening' – until it fails in a brittle manner and breaks in two.

8.36 This scanning electron micrograph shows the fracture surface of a cobalt–chromium–molybdenum gas turbine blade from a jet engine.

Study of the fracture faces of a specimen like this enables the scientist to trace a crack to its origin. The dark grey surface in the picture contains 'river markings', so called because they join like river tributaries in the direction in which the crack is propagating. Thus the crack in this specimen has moved from right to left. The lines on the oblique surface in the top right corner are the wave-like formations created when a crack weaves up and down, as described above.
SEM, ×200

8.37 The effects of a material's atomic structure can be evident even in failure. In this specimen of the ceramic alpha alumina, a crack moving from right to left has been driven downwards. As it moved, it followed the planes of lowest energy – those which fracture most easily – in the ceramic's atomic lattice structure. In this instance, two low-energy planes lie at right angles to each other, so that the crack has propagated downwards in a series of steps. The 'steps' stop abruptly at left where the crack plunges sharply downwards at a grain boundary.
SEM, ×2700

8.38 It is not only metals and specialist ceramics whose fracturing is studied microscopically. The 'face' in the micrograph opposite is a fracture feature in a piece of the common plastic material, polystyrene. Polystyrene is prone to reduction in the atomic weight of its long chain molecules. A year's exposure to desert sunshine is sufficient to halve its molecular weight, with an attendant loss of strength, due to photooxidation – a combined attack of ultraviolet light and atmospheric oxygen.
SEM, ×8250

QUANTITATIVE MICROSCOPY

As engineers have exploited materials more widely, they have tended to work closer to the limits of the materials' capabilities. In order to do so safely, both mechanical testing and microscopy have had to ensure that the properties of a given material are well understood and predictable. The desire for more precise information has led to a significant increase in the use of *quantitative* microscopy, where actual values are measured for observed properties and microstructures. Quantitative microscopy takes many forms, from the simple measurement of a material's hardness to the detailed mapping in a scanning electron microscope of its constituent elements. Elemental mapping relies on the X-rays emitted when the electron beam hits a specimen; each element emits X-rays of a different wavelength, providing a chemical signature.

8.39 The most common test in quantitative microscopy is the microhardness measurement of a material. A pyramid-shaped diamond, with the point facing downwards, is glued to a special, movable, front element of a microscope's objective lens. Viewing the specimen through the microscope, the investigator can see well enough around the diamond to select in a pair of cross-hairs the exact part of the specimen he wishes to test. A known weight then drives the front of the objective onto the material, where the diamond produces a tiny indent, a few micrometres across. Measurement of the indent gives a precise value of the material's hardness at that spot. The picture shows a series of microhardness indents in a metal surface.

LM, interference contrast, polished section, ×480

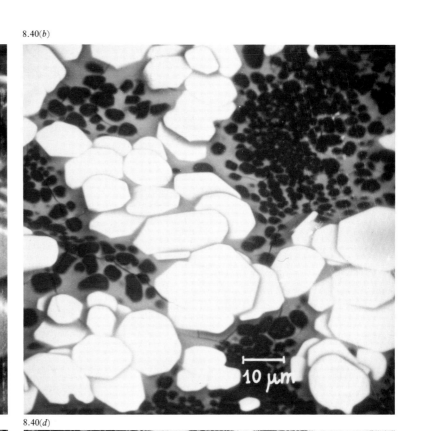

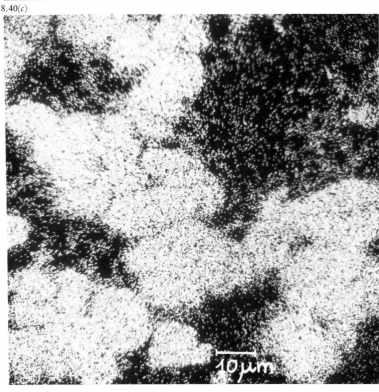

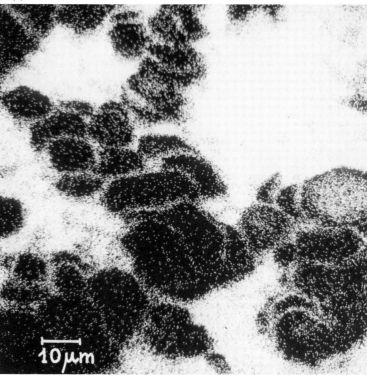

8.40 This group of scanning electron micrographs shows the same oxidised area of a silicon nitride ceramic viewed by four different imaging and analysis techniques. Figure 8.40(a) is a conventional secondary electron image which shows that the oxidised surface consists of polygonal plates set in a somewhat granular background matrix, but it gives no information about the composition of these structures. Figure 8.40(b) is an image formed from back-scattered electrons, and it tells the microscopist that the brightest areas – the white polygons – contain an element of a high atomic number. This is because heavier elements back-scatter more electrons.

Figure 8.40(c) is a map of X-rays emitted at the wavelength for the metal cerium, and it shows that the polygonal plates consist largely of cerium oxide, which was used in making the ceramic. Figure 8.40(d) is a second X-ray map, at the wavelength for silicon; it shows that the granular background matrix is predominantly silicon. X-ray mapping can be used to identify all elements with an atomic weight greater than that of fluorine.

SEMs; 8.40(a) secondary electron mode; 8.40(b) back-scattered mode; 8.40(c) cerium X-ray map; 8.40(d) silicon X-ray map; all ×1000

CHAPTER 9
EVERYDAY WORLD

IN this final chapter, the subjects are close to home: familiar items seen in unfamiliar close-up. The pictures illustrate well the fascination which microscopes hold for those lucky enough to use them. A single image can reveal the structure or workings of an object which would take hundreds of words to describe with equal clarity. The most mundane objects can take on a startling or striking appearance.

9.1 This picture is of the most ordinary subject it is possible to imagine – house dust. The specimen was obtained simply by tipping out the contents of a domestic vacuum cleaner. House dust consists of fragments of soil brought in on the soles of feet or shoes, fibres which have escaped from clothing or hair, and skin scales from the inhabitants of the house. Not all the components of dust are dead, however; in this scanning electron micrograph, the object on the right-hand side of the picture is a tiny dust mite, a species of *Glycyphagus*. All vacuum cleaners contain dust mites, and so do all carpets and mattresses. Each gram of dust contains about 1000 mites. They spend their days perambulating a landscape of, to them, huge rocks and tangled jungles. They eat the human skin scales with which we fill their microscopic world. This one is moving from right to left. The boulder it is about to encounter is a tiny grain of sand.
SEM, ×145

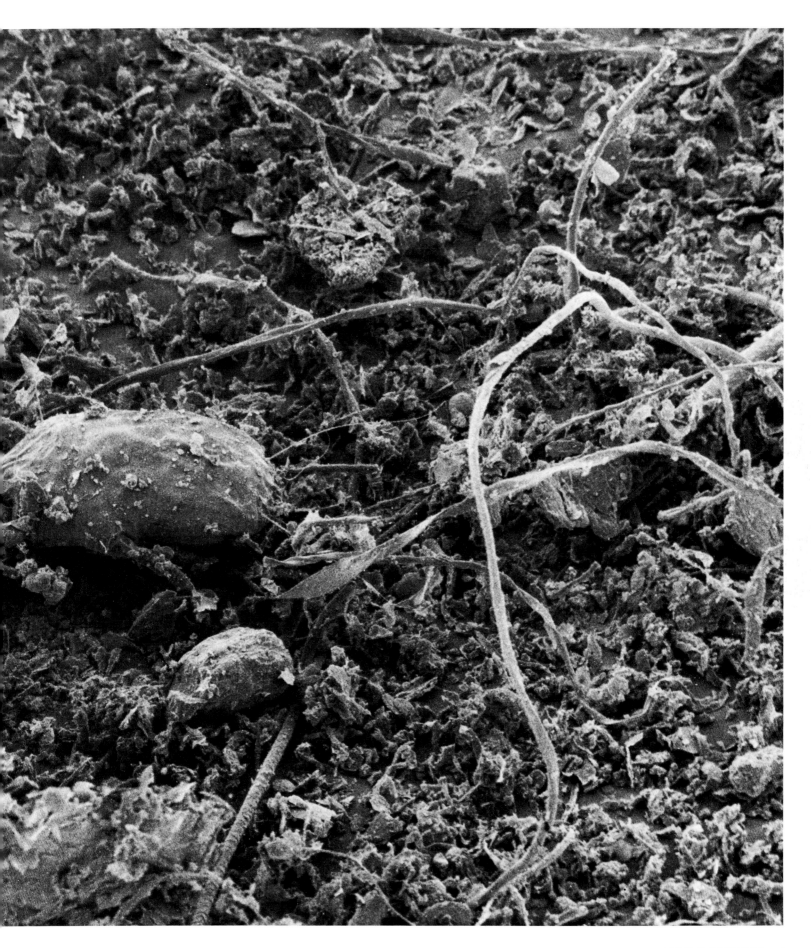

FABRICS

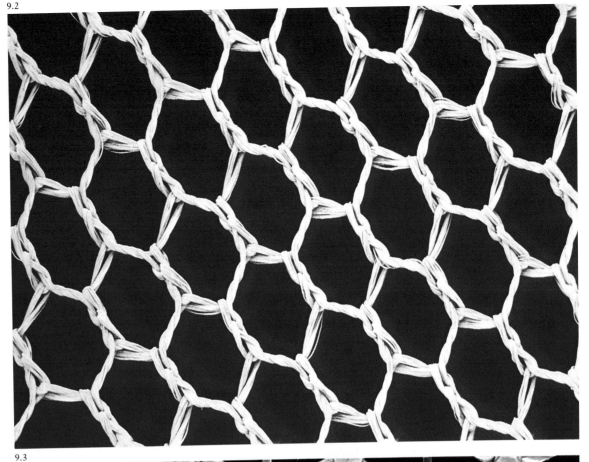

Fabrics are produced from a variety of fibres, both natural and man-made. Natural fibres come from plants (cotton, flax), animals (wools), or insects (silk). With the exception of silk, natural fibres are usually short; they are also very fine, and must be spun together to produce a thread of useful length and thickness. Man-made fibres, on the other hand, are extruded continuously from dies; their length and thickness is controlled during manufacture. The fabric itself is made by weaving or knotting threads together.

9.2 Lace-making is a skilled craft; in this micrograph of a piece of machine-made lace net, each cell of the net can be seen to be a complex series of knots. The thread is a multi-stranded cotton and polyester yarn.
SEM, ×7

9.3 This fabric is machine-made from an artificial fibre. It is a piece of net made from polythene. Each cell of the net is identical to its neighbours, and the thread is a single extruded fibre of plastic. The micrograph was made using polarised light, and this accounts for the colours, which correspond to residual stresses in the plastic material of the fibre.
LM, polarised light, ×35

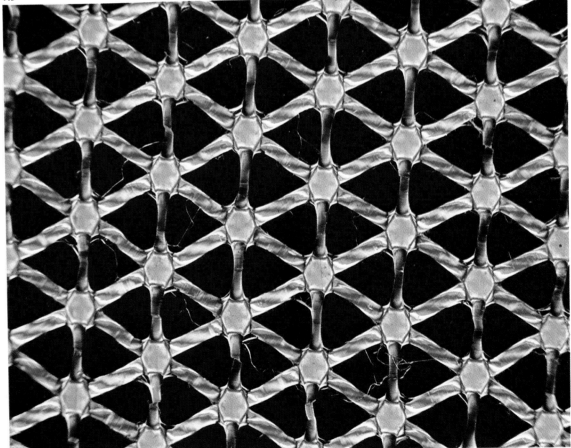

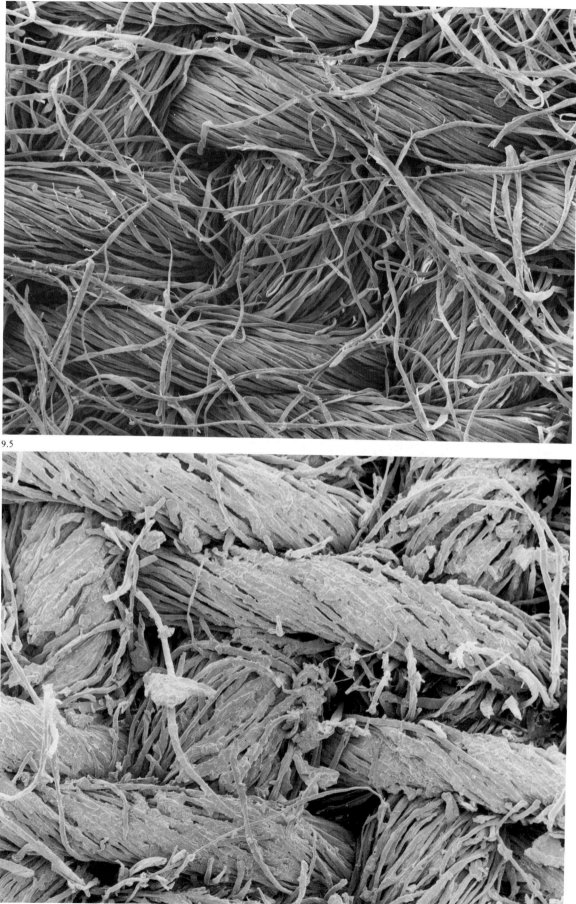

9.4–9.5 This pair of scanning electron micrographs shows the difference between a clean and dirty shirt collar. In Figure 9.4, the clean collar is seen to be made of woven cotton threads, each of which consists of many individual cotton fibres spun together. Figure 9.5 reveals the transformation wrought by just one day's city wear. The fabric has become encrusted with a layer of grease, sweat and particles of dust, together with dead scales from the wearer's skin. Happily, washing the fabric in detergent will restore the pristine appearance shown in Figure 9.4.

SEMs, both ×100

9.5

9.7

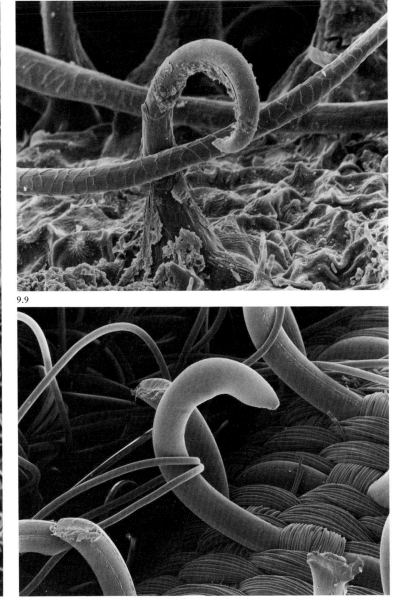

9.8

9.9

VELCRO

One day in 1955, the Swiss inventor George de Mestral was out hunting in the Alps with his dog. Both became covered with burrs from the burdock plant. As he picked the offending objects from his clothing and from his dog's coat, Monsieur de Mestral had a bright idea. Why not design a fastener using the same principle? Velcro had arrived.

Velcro is made from nylon and

it comes in two parts – one side consisting of hooks and the other side consisting of loops. When the two parts are pressed together, the hooks engage the loops, and the fastening is made. Velcro is strong enough to resist a direct pull, but is easily dismantled by a peeling action. Its particular virtue is its ease of use, even when manual skill is impaired through injury or by wearing gloves. Velcro fasteners have been to the top of Everest, and out into space. It is also a durable fastening, able to withstand laundry, and able to work after 5000 cycles of opening and closing.

9.6 This photograph shows the two parts of Velcro in the separated state. The hooks (top) consist of loops of thick nylon which have been snipped open during manufacture. The loops (bottom) are closed, and are made from multiple strands of thinner nylon which can pass through the gaps in the hooks.
Macrophotograph, ×85

9.7 Plants have been using the idea of Velcro for millions of years. This scanning electron micrograph shows a fruit of the goosegrass, *Galium aparine*, attached to a woollen sweater. The loose fibres on the surface of the garment have entangled themselves in the projecting hooks on the surface of the fruit.
SEM, ×20

9.8 At higher magnification, the mechanism is clearer. Each hook is beautifully curved and finishes in a sharp point. In this picture, one hook has captured a single fibre of wool; another fibre is in the background. The scales on the surface of the wool are characteristic of animal hair; in this case, that of sheep.
SEM, ×250

9.9 The equivalent view of Velcro shows that the stout hooks are woven into a base of fine nylon material for flexibility. Nylon is extruded from fine dies, and shows no surface texture. The hook in this picture is about eight times larger than the natural one shown in Figure 9.8.
SEM, ×30

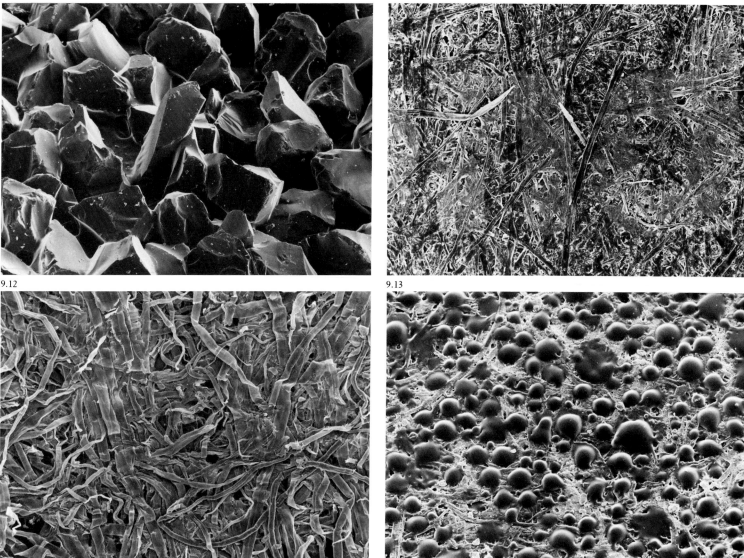

9.10

9.11

9.12

9.13

PAPER

Most paper is made from cellulose fibres extracted from wood. Other fibre sources such as rags, flax and hemp are used to make fine quality papers used for legal documents or cigarette wrappings. In the production process, the fibre source – wood chips, cleaned rags, the stems of flax, jute, or bamboo – is beaten in water to make a pulp. The liquid pulp is treated with chemicals to remove impurities, bleached, and filtered on a woven screen, producing a

wet mass of tangled fibres which is then rolled, compressed and dried to yield the sheet of paper.

Paper varies enormously in quality, according to its thickness, density and fibre source. The cellulose fibres from softwoods (conifers) are longer than those from hardwoods; softwood paper is stronger, but hardwood paper has greater opacity, and can be made smoother. Cotton fibres from rags are very long and low in impurities; the paper they produce has great durability and quality, and is used for banknotes, tracing paper and carbon paper.

9.10 The paper base of sandpaper is a high-quality product made from hemp

or jute fibres. In this scanning electron micrograph, the paper itself is not visible. The picture shows the particles of abrasive, consisting of ground glass, which are bonded to the surface of the sheet. The specimen is a coarse grade of sandpaper; finer grades use smaller glass particles.
SEM, ×50

9.11 Newsprint is made from sawn logs ground between abrasive stones. The pulp contains all the impurities of the original timber: broken fibres, lignin and other cell wall components. Paper from this type of pulp is never very white, and discolours easily. It is opaque, however, and accepts printing inks well. The specimen is from a page of *The Times* newspaper, and shows the letters 'nd' – from the word 'London' – printed on it.
SEM, ×60

9.12 Viewed at roughly the same magnification, soft toilet tissue has a much more open structure, with fewer impurities and longer, wider fibres. It is made from pulp which has been beaten and chemically refined. As a result, it is very absorbent.
SEM, ×50

9.13 Each of the round objects in this micrograph is a tiny resin sphere filled with glue and incorporated into the surface of a sheet of paper. When the sheet is pressed onto a firm surface, it becomes stuck as some of the spheres burst and release their adhesive. The process can be repeated many times, since only a few of the spheres burst on each occasion. The picture is of a Post-It sticker.
SEM, ×70

WATCH

9.14 The Swiss watch epitomises precision engineering in most people's minds. It is a reputation that is well deserved, as this scanning electron micrograph shows. Even when substantially magnified, the mechanism presents a picture of precision; the slightly off-centre appearance of the middle gear wheel is in fact due to the specimen being tilted on the microscope stage. The picture is of the crown wheel of a 17-jewel Incabloc movement. When the watch is wound, the winding pinion (the small gear wheel seen edge-on at bottom centre) rotates, meshing with the crown wheel. The rotation of the crown wheel causes the ratchet wheel at top left to turn, and this wheel is attached to the barrel which holds the mainspring of the watch. The mechanism is very clean; the very small particles of dust on the surface and on some of the teeth of the crown wheel would not affect its function at all. The watch had never been serviced since its assembly, and this is reflected in the absence of screwdriver marks in the slotted pillar of the crown wheel.

SEM, ×16

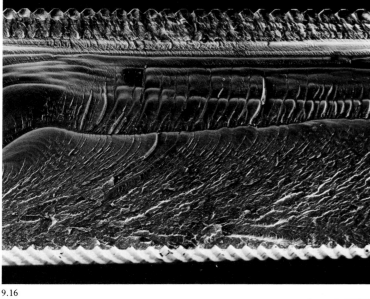

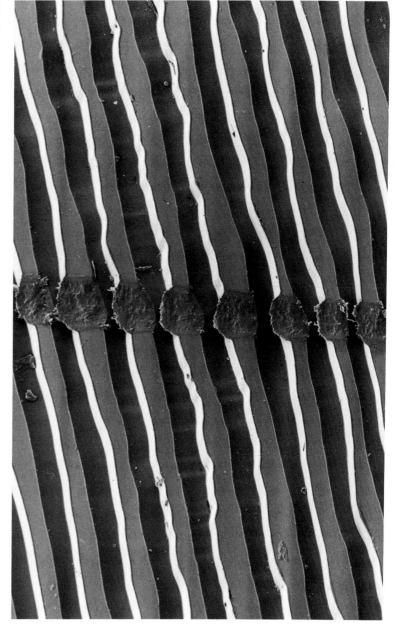

RECORDS & DISCS

A gramophone record is a remarkable object, even though a very familiar one. Within its plastic surface is stored all the information to reproduce 30–40 minutes of the sound made by an orchestra consisting of dozens of instruments. Microscopy shows how this is achieved, and in particular it emphasises the difference between the old-fashioned long-playing record and its modern counterpart, the compact disc.

9.15 An LP has two 'sides'. Each consists of a modulated spiral groove pressed into the surface of the plastic disc. In this scanning electron micrograph of a cross-section through a broken LP, the grooves appear as the saw-tooth patterning towards the top and bottom of the picture. The markings on the main body of the disc are fracture patterns in the PVC from which the record is made.
SEM, ×30

9.16 Each V-shaped groove on the LP's surface has undulating side walls. The stylus follows the contours of these waves, and the way in which it moves determines the sound which is produced by the loudspeakers. In a loud passage, for example, the waves on the sides of the groove are deep; in a quiet passage, they are shallow. A high note is recorded as a rapidly changing wave, whereas a note of low pitch is recorded as a slowly changing wave. In this micrograph, a diamond stylus sits in the groove. Before making the micrograph, the stylus was carefully cleaned; nonetheless, the picture shows that it is very dirty. The small particle in front of the stylus is a piece of dust. As it passes over the dust, the stylus will be rapidly jerked upwards and this will give rise to a high-pitched click in the loudspeakers. The part of the groove in this picture represents a loud orchestral passage in the *Sinfonietta* by Janacek.
SEM, ×90

9.17 A major drawback with LPs is the ease with which they can become scratched. This micrograph shows the effect of accidentally drawing a fingernail across the surface of an LP. The resulting scratch narrows the top of each turn of the groove, and causes a sudden movement of the passing stylus. The result is an annoying, loud click each time the disc revolves – every 1.8 seconds.
SEM, ×60

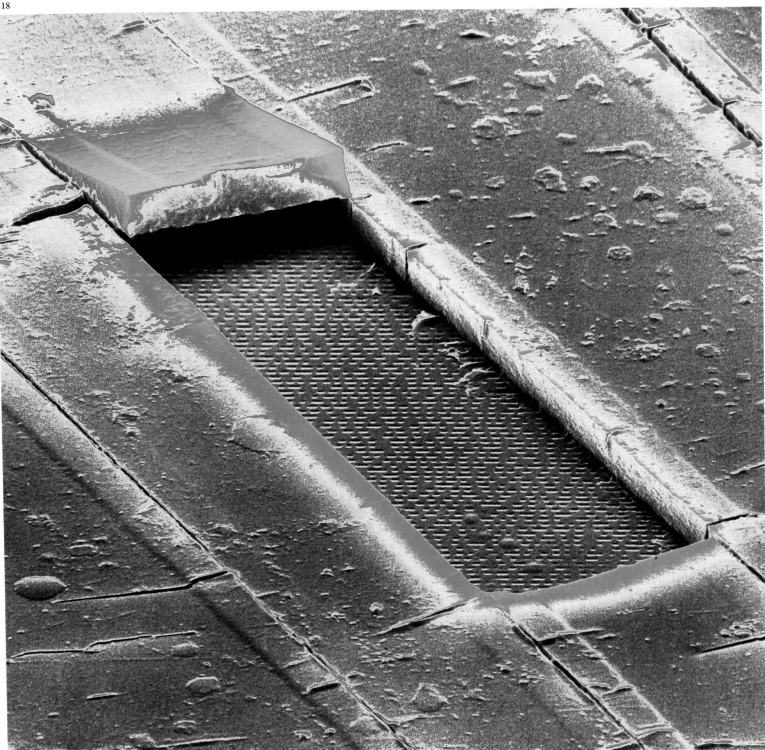

9.18 The compact disc works in a completely different way to the LP. One side of the disc is pressed to produce a pattern of very small elongated bumps, laid out in the form of a fine, continuous spiral which, in a typical CD, is 20 km long. The turns of the spiral are so close together that 60 of them would fit into the width of a single groove of an LP. The size of each bump and the rate at which adjacent bumps change in size determine the volume and pitch of the sound produced. The bumps are not touched by a stylus. Instead, they are coated with a thin layer of aluminium, which reflects the light from a laser beam focused on the metalised surface. As the disc rotates, the light is reflected to a sensor as a stream of rapid flashes, and this signal is processed to produce the music in the loudspeakers. The delicate surface of the disc is protected by a layer of transparent plastic. In this picture, the protective layer has been cracked open and partly removed. Beneath the protective layer, the pattern of bumps is visible – part of the first movement of Mozart's 40th symphony. The bumps vary in length from 0.83 micrometres to 3.56 micrometres. These distances are close to the wavelength of visible light. This is why shining a light onto a CD produces a rainbow-coloured reflection, by the process of diffraction.
SEM, false colour, ×1040

FOOD

Cooking can be a chore or an art form – depending on the cook. To a microscopist, however, cooking rarely improves a raw ingredient. The exquisite architecture of cells and tissues is irreversibly destroyed as heat coagulates proteins, dissolves fats, and hydrolyses fibres and starch. These, of course, are the very changes which make food pleasant to taste and easy to digest. And most of us, after all, cook with eating in mind, not microscopy.

9.19 Meat is the muscle tissue of animals. In this piece of roast beef, the muscle fibres can be seen as the rectangular rods running from top left to bottom right. The striations along each rod correspond to the arrangement of actin and myosin filaments in the living muscle. The entire specimen is coated with a thin layer of fat released by the cooking. SEM, ×370

9.20–9.21 The potato is a rich source of energy in our diet due to the presence of starch in its cells. In the slice of a raw (living) potato in Figure 9.20, the starch appears in the form of particles called amyloplasts, groups of which can be seen looking like eggs in a nest. Other cells in the picture are not cut through the centre, and appear to contain no starch. The picture is full of the beauty and detail which typifies micrographs of living biological materials. After 20 minutes in boiling water, by contrast, the organisation of the potato has changed completely (Figure 9.21). The outline of the cells is still visible, but the starch has been converted into a bulky, glue-like mass. This is more easily digestible, but less visually pleasing. SEMs, both ×170

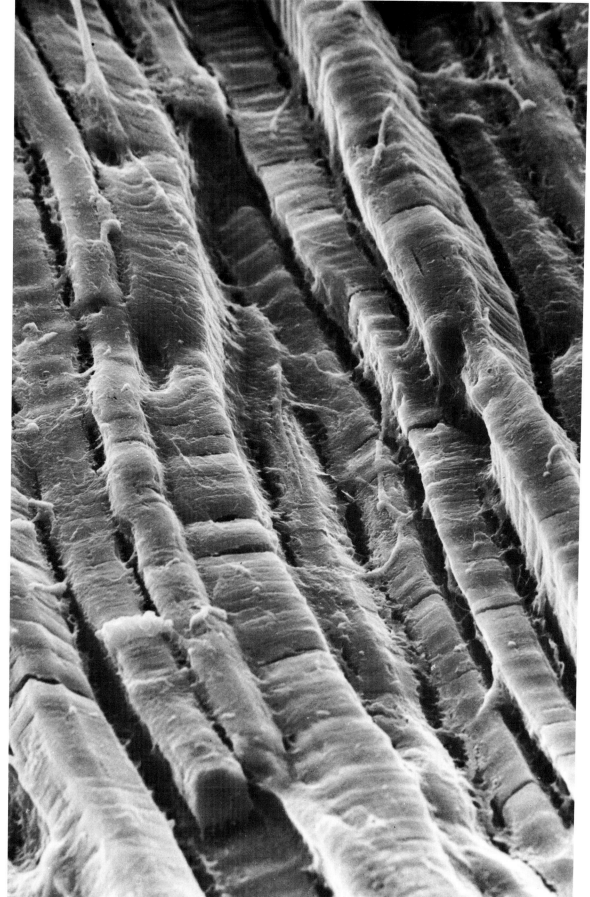

9.20

9.21

TECHNICAL APPENDIX

THE HISTORY OF MICROSCOPY

The word microscope was coined in 1625 by Giovanni Faber. He used it in a letter describing an invention by Galileo. The instrument to which it applied was not what we would recognise today as a microscope; instead, it was a telescope adjusted so as to view nearby objects. The credit for the invention of the compound microscope must go to a Dutchman, Zacharias Jansen (1588–1630). His microscope consisted of two positive lenses and was produced in 1608. This design has persisted to the present day, although some of the most striking discoveries in the early years of microscopy did not involve its use.

The first book of microscopical results to be published was by Robert Hooke (1635–1703). Hooke was curator of experiments at the Royal Society of London, which was founded in 1660. In 1663 he was asked to deliver a weekly series of lectures on his experiments with microscopes. It was an auspicious start for the science

of microscopy. In only his second demonstration, Hooke decided to look at the structure of cork. By cutting thin sections of the material with a razor blade, he was able to discern the tiny compartments of which it is made, and he called them 'cells'. In a single experiment he had invented the fundamental technique of the microscopist, and named the fundamental unit of life.

His results were published in a book called *Micrographia* in 1665. Hooke's specimens ranged from the point of a needle to the compound eyes of insects, and he was quick to realise that a huge gulf existed between the works of man and those of nature. Of man-made things he wrote: 'If examined by an organ more acute than [sic] that by which they were made, the more we see of their shape, the less we see of their beauty; whereas in the works of Nature, the deepest discoveries show the greatest excellencies.'

Micrographia contains details of the microscopes which Hooke used, and those which he tried and rejected. His favoured microscope had two lenses, corresponding to the objective and eyepiece of the modern compound microscope. But he suggested that the very best results would probably be obtained by using a single lens. He even gave details of how such a microscope could be made, but concluded that it would be inconvenient in use compared to his two-lens design. He was right: to obtain high magnification from a single

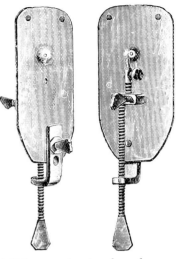

Antoni van Leeuwenhoek. The picture was first published in his book, Arcana Naturae detecta, *in 1695.*

lens, the distance between the lens and the object (the working distance) has to be extremely short.

Three years after *Micrographia* was published, a Dutch draper visited London. His name was Antoni van Leeuwenhoek (1632–1723), and he has become known as the father of microbiology. During his lifetime he built over 500 microscopes of the single-lens type that Hooke had described and rejected as impractical. With their aid he made the first descriptions of micro-organisms, including bacteria. He was a dedicated scientist, and wrote regularly to the Royal Society describing his discoveries until his death at the age of 90. He was surprisingly reticent, however, about demonstrating his discoveries. His numerous visitors were always shown routine specimens such as fleas or flies. There is no record of anyone being shown exactly how it was that he was able to see bacteria. But his drawings are unambiguous, and a testament to his skill as an observer. His demonstration of previously unknown 'animalcules' represents a milestone in the history of biology and medicine.

The large number of microscopes van Leeuwenhoek built reflected a curious attitude; each was designed and used for more or less one specimen only, which was impaled on a pin, and

A 'blue fly', from Robert Hook's Micrographia, *published in 1665. Although not greatly magnified, this drawing is the earliest clear representation of the compound eye of an insect.*

A drawing of one of Robert Hooke's microscopes, from Micrographia. *Light was provided by an oil lamp, and concentrated on the specimen by means of a glass globe filled with water.*

A 19th century drawing of one of van Leeuwenhoek's microscopes. The drawing was made by John Mayall, a secretary of the Royal Society of London. The lens was only a few millimetres in diameter and the whole instrument, which was held in the hand, was about 5 centimetres high.

was almost an integral part of the microscope itself. His most powerful lenses could magnify about 300 times and had a resolution of about 1.4 micrometres.

The single-lens microscope was developed further in Italy and Holland, but it is as the screw-barrel microscope of James Wilson of England that it was best known. Produced in 1702, it had a screw-barrel focusing mechanism, together with a condenser lens. Such instruments were widely bought by keen amateurs, who were also supplied with specimens that typically included sectioned plant tissue, the human louse, human hair, and skin scales of fish such as the sole.

With hindsight, it is clear that Hooke's assessment of the single-lens microscope was correct. It was not convenient to use, and it is therefore all the more to van Leeuwenhoek's credit that he made so valuable a contribution to microscopy. He did not develop the instrument; he demonstrated the power of its use. Hooke had taken common objects and described their magnified appearance. Van Leeuwenhoek described the

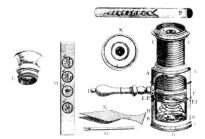

James Wilson's screw barrel microscope of 1720 was a development of van Leeuwenhoek's simple design. In use, the handle was gripped and the whole instrument pointed towards a light source. The light entered through a condenser lens, and the specimen, at EE, was viewed through the single objective at G. The screw barrel mechanism was a refinement for focusing. The picture also shows the four-specimen ivory slider, M, which was supplied as part of the kit.

appearance of things which had never been known to exist before.

In the half century following *Micrographia*'s publication, microscopy became firmly established. Because techniques for the observation of minerals, metals and other inorganic objects had not yet been developed, the emphasis was on biological specimens. But within the limitations imposed by having to view whole organisms or crude, hand-produced sections, many remarkable observations were made. Van Leeuwenhoek described red blood cells and spermatozoa as well as bacteria. His countryman Jan Swammerdam (1637–1680) founded the science of invertebrate anatomy with brilliant dissections of insects. In Italy, Marcello Malpighi (1628–1694) observed the circulation of blood in a frog's lung using an injection technique devised by Christopher Wren.

Compared to this fertile early period, the 18th century lacked accomplished microscopists. Progress was mainly concerned with the microscope itself. One invention in particular did much to widen the instrument's scope. This was the solar microscope of Nathaniel Lieberkühn (1711–1756). Its innovative feature was a concave spherical mirror fitted around the viewing lens. This device, called a Lieberkühn to this day, enabled opaque objects to be lit from above by reflection of sunlight. It liberated microscopy from having to use translucent specimens. Lieberkühn himself used the device to

study the mucous membrane of the alimentary canal, but the way was also open to examine objects of industrial importance. In 1770, a Dr Hill published *The Construction Of Timber Explained By The Microscope*. This work caused a sensation; it suggested for the first time that a link existed between the structure of an object and its function. Hill showed that different types of wood, useful for different purposes, had correspondingly different structures under the microscope. This idea is so integral to modern thought that it is hard to imagine the effect of its first expression.

At the beginning of the 18th century the microscope was still a crude device, both optically and mechanically. The tube holding the lenses, for example, was often made of cardboard. The first move towards precision construction was made in 1744 by John Cuff (1708–1772). His microscope was made of brass, and based upon a famous earlier design by Edmund Culpepper (1660–1740). Another feature which appeared at this time, and which is still in universal use, was the rotating nosepiece with two or more objective lenses of different magnification. This made its debut in 1746 on George Adams' New

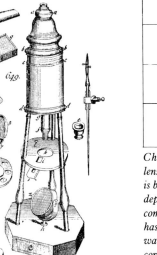

The Culpepper microscope of 1730, in a version by Edward Scarlett (1677–1743). This two-lens design featured a tripod stand which allowed an integral mirror to be fitted beneath the specimen stage. A wooden base contained a drawer for storing accessories and specimen holders, which in this instrument could accommodate eight specimens.

Universal Microscope. These improvements did not materially alter the poor image quality of the microscopes of the period, nor did they stimulate imaginative use of the instrument. The century was summed up in 1849 by John Quekett in his famous book, *Practical Treatise on the Use of the Microscope*: 'The discoveries at this time were few and comparatively unimportant, and little or nothing more was exhibited . . . than the objects contained in the ivory sliders, with which all the above microscopes were supplied; and he who could exhibit these objects well was considered a proficient in the art.'

The reason for the lack of good observation during the 18th century was undoubtedly the poor quality of the optics available. Lens makers assumed that the only important property of the starting material was its transparency; if you could see through it, you could grind a lens from it. But simple glass lenses suffer from two major defects. The first is spherical aberration, which is a function of the shape of the lens surface. The second is chromatic aberration, which is caused by the way in which glass refracts light. Light of different wavelengths is refracted to different extents by a given piece of

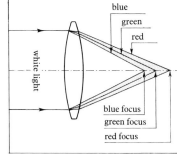

Chromatic aberration in a simple glass lens. White light passing through the lens is brought to focus in different planes depending upon the wavelength of its component colours. This is because glass has a different refractive index for each wavelength of light. The defect is corrected in compound lenses by using diverging elements of a different glass.

glass (an effect known as dispersion). This means that a simple lens will always produce an image which is blurred by the presence of coloured fringes around the edge of any well-defined object.

Achromatic lenses that partially corrected this defect were already used in telescopes. The first person to use

such a lens in a microscope was Benjamin Martin (1704–1782), an English schoolteacher. His projection microscope was not much more than an entertainment; it threw an enlarged picture of the object (a flea was popular) onto the wall of a darkened room. The lens contained both positive and negative elements – a so-called compound lens of the triplet type. Initially these lenses were designed by trial and error, and it was not until 1820 that a serious mathematical approach to the problem was adopted, pioneered by J.J. Lister (1786–1869).

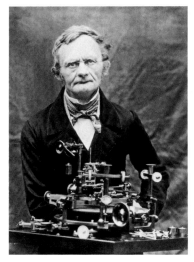

Friedrich Nobert, photographed with his ruling engine in 1867.

One of the difficulties with assessing lens performance was the lack of a standard test object. It is obviously futile to try to obtain a clearer image of something if its real structure is unknown, or variable. A convenient natural grating of fine lines is to be found on the wing scales of butterflies, and these were used as a test object for microscopy by the Edinburgh doctor, C.J. Goring. Better still are manufactured diffraction gratings – glass slides with parallel lines ruled on them at known spacings. In 1845, the German instrument maker Friedrich Nobert (1806–1881) began to produce such gratings. He used a circle-dividing machine of the type commonly employed to draw angular scales for astronomical instruments, to which he fitted a diamond scribing point attached to a reduction lever. He was able to rule lines which were separated by as little as 0.11 micrometres, with an accuracy better than 10 per cent. Unknown to his contemporaries, such a fine spacing cannot be resolved by a light microscope; the first image of Nobert's finest lines was only produced in 1966 by an electron microscope. A former secretary of the Royal Society, John Mayall, wrote in 1885: 'When the history of mechanical inventions of our time comes to be written, a large measure of credit will be assigned to the mechanical genius of Herr Nobert, embodied in this ruling machine.'

Nobert produced a series of plates with spacings from 2.32 down to 0.11 micrometres. Microscope manufacturers competed with each other to resolve the finest lines, and the result was that the latter part of the 19th century can justly be regarded as the golden age of light microscopy. Instruments of extreme precision and great versatility were its hallmark. The London manufacturers Andrew Ross and Hugh Powell were outstandingly

The Powell and Lealand No. 1 microscope was manufactured for many years during the second half of the 19th century. It was a precision piece of equipment, made in brass.

successful, and the Powell and Lealand microscope is still considered by some to be the finest instrument of its type ever made. Features which became standard during this period included the rotating specimen stage, the substage condenser within a centering mount, adjustable draw tube length, and spring-loaded fine-focus bearings.

The achievements of observers during the 19th century were also significant. In biology, two books in particular represented landmarks. In 1839, the Prussian physiologist Theodore Schwann (1810–1882) published *Microscopic Investigations on the Accordance in the Structure and Growth of Plants and Animals*, in which he identified the cell as the basic unit of life. In 1858, the German pathologist Rudolf Virchow (1821–1902) published *Cellular Pathology Based on Physiological and Pathological History*. This established the concept that disease affects the functioning of cells, and placed histology firmly at the

forefront of medical investigation. Virchow also coined the aphorism *Omnis cellula e cellula* ('All cells arise from other cells'). This idea, which seems banal today, was of incalculable importance in its time. It explained not only how a single cell could grow into an embryo, then an adult, but also how disease organisms could increase and spread within a host tissue.

The physical sciences also blossomed during this period. In 1828, Edinburgh geologist William Nicol (1768–1851) invented the Nicol prism, which allowed the properties of crystals to be studied using polarised light. This led to the production of microscopes with calibrated specimen stages; the era of quantitative microscopy had begun. The microscope also maintained its place as a drawing-room entertainment. The favourite microscopical amusement of the Victorian period was the arrangement of diatoms in geometrical or other patterns.

The use of photography to record microscope images preceded Daguerre's public announcement of the photographic process in January 1839. Henry Fox Talbot (1800–1877) took photographs at ×17 magnification with his solar microscope while perfecting his own photographic method. When Talbot presented his work to the Royal Society four weeks after Daguerre's announcement, he predicted the major role photomicrography would play in the emerging sciences of bacteriology and metallurgy. Photography and microscopy have been intimately linked ever since.

The early 19th century was a period when microscopes improved dramatically with each new model, and manufacturers competed fiercely. It became clear to one man, however, that whatever the excellence of the construction of the microscope, there was a limit to its ability to resolve fine details. Ernst Abbé (1840–1905) was the son of a German textile spinner. In 1866, he joined the newly formed company of Carl Zeiss, in Jena. He believed in designing lenses according to the principles of physics, and he was aware that the types of glass then available did not have the properties necessary for full correction of chromatic aberration throughout the visible spectrum. Together with the glass-maker Otto Schott (1851–1935), he set about rectifying this deficiency. By 1886, 44 types of optical glass were

Ernst Abbé, who brought the principles of physics to bear on the design and manufacture of lenses.

available in Jena. They varied in their refractive index, and also in their dispersion – the variation of refractive index with wavelength. Jena glasses enabled lenses to be manufactured with colour correction at three different points in the spectrum. Today these lenses are known as apochromatic.

Abbé also showed that in order to obtain the best possible resolution from a lens, it had to be able to collect diffracted rays of light as they left the specimen. He coined the term numerical aperture to express this ability in a mathematical form. In simple terms, it means that for a given magnification, the larger the front glass of the lens, the better. The light gathering ability of a lens is improved if it is operated in a medium of high refractive index, and Abbé designed his high-powered objectives to be used with oil bridging the gap between the specimen and the front of the lens.

The 20th century has been one of greatly increased convenience for the optical microscopist. Light microscopes are widely used in research, medicine and industry, and in the vast majority of applications the operator wants a tool which is reliable, quick and simple to work. Most modern light microscopes are built on modular lines, with interchangeable illumination systems, devices for specimen manipulation, and automatic recording features, whether by photography or by electronic techniques which allow computerised image analysis.

Classical microscopy was mainly concerned with examination of thin specimens whose contrast was enhanced by staining. The techniques involved remain invaluable in modern

histology and cell biology. However, the 20th century has produced novel ways of increasing image contrast. The first was the phase contrast technique developed in 1942 by Nobel laureate Frits Zernicke (1888–1966), a Dutchman. More recently, a variety of interference techniques have been developed which are an improvement on phase contrast because they can be used to make quantitative measurements.

The ultraviolet microscope designed and built at the turn of the century by two Zeiss employees, August Köhler and Moritz van Rohr, never became popular. It demonstrated that using short-wavelength light improves a microscope's resolving power, but it was inconvenient to use and expensive. However, it was the forerunner of the ultraviolet fluorescence microscope, which is a very powerful modern tool.

A modern light microscope in use. Various features designed for operator convenience can be seen, such as the inclined binocular viewing head, the rotating nosepiece holding four different objectives and, in the background, an automatic exposure unit to control the camera which is mounted on top of the microscope.

Combined with sophisticated staining methods, the fluorescence microscope makes it possible to determine the positions of molecules within cells – molecules far too small to be visible in the conventional light microscope.

Light is a form of wave motion, and because of this it cannot resolve the distance between two objects if that distance is less than roughly the wavelength of the light itself. Using short-wavelength ultraviolet light offers only a minor improvement. As far back as the latter part of the 19th century, Abbé realised that the only way forward was to use radiation of a much shorter wavelength. What was needed was a different sort of microscope altogether. The fulfilment of this dream has been the major achievement of 20th century microscopy. Two related instruments of enormous power have been developed: the transmission electron microscope (TEM) and the scanning electron microscope (SEM).

The electron was discovered in 1897 by the Cambridge physicist J.J. Thomson (1856–1940), who regarded it as a particle. Others considered that it had wave-like properties, and in 1924 Louis de Broglie (1892–1966) demonstrated that a beam of electrons travelling in a vacuum behaves as a form of radiation of very short wavelength. In the same year, the first electron lens was built by Gabor in Berlin, but neither he nor de Broglie made the conceptual leap of envisaging a microscope using electrons as the imaging radiation. Two years later, Hans Busch, also working in Berlin, published a now famous explanation of how the electron lens worked, but he too failed to make the leap. That distinction was left to two other Berlin workers, Max Knoll and Ernst Ruska. In 1931, they produced a picture of a platinum grid magnified 17 times, using electrons as the imaging radiation. Knoll left the team shortly afterwards, and it fell to Ruska (b. 1906), working alone as a Ph.D. student, to produce the first electron microscope which surpassed the light microscope in resolving power. In 1933, he demonstrated a point-to-point resolution of 50 nanometres in a specimen of cotton fibres. This improved by a factor of 4 the resolving power of the optical

The first TEM, built in Berlin in 1931. The microscope column is on the left; the main bulk of the instrument is the power supply which generated the high voltage necessary to accelerate the electrons.

The first Cambridge SEM, in 1953. The column is dwarfed by the power supplies. The vacuum pumps can be seen on the floor at bottom left, beneath the base of the column. Curiously, the display screen – in the form of a television set – has been inconveniently located half way up the wall in the corner.

Ernst Ruska (left) and Max Knoll, photographed in 1932 in Berlin.

microscope using visible light. The image in Ruska's microscope was formed as a result of electrons passing through the specimen. It was the world's first TEM.

It was over half a century later, in 1986, that Ruska's achievement was finally rewarded with the Nobel prize for physics. At the time, his pioneering work was greeted with a mixture of derision and hostility. The first electron microscope pictures gave no clue of the revelations which were to come – they showed the badly charred remains of the specimen! It was even suggested by certain professors that there were no permanent biological structures smaller than those which could be seen with the light microscope. Undaunted, Ruska continued his work. It was quickly realised that specimen damage was due to absorption of energy from the electron beam, and that this could be reduced with thinner specimens and higher accelerating voltages. In 1936, Driest and Müller photographed the wing of a housefly in an undamaged state using an electron microscope.

The first commercial TEM was built in Britain by Metropolitan Vickers in 1936, followed in 1939 by the German firm of Siemens, whose design has been copied repeatedly by other manufacturers. The Second World War arrested electron microscope development in Europe, but not in the United States, where a resolution of 1 nanometre was demonstrated in 1946 by J. Hillier at RCA. A modern TEM can achieve resolutions of 0.2 nanometres, one thousand times better than the light microscope.

The TEM relies on electrons passing through the specimen in order to produce the image. Such a specimen has to be very thin – of the order of 100 nanometres (1/10 000th of a millimetre) or less for a typical TEM working at 60 000 volts. One consequence is that it is very difficult to study the three-dimensional relationships within a specimen by transmission electron microscopy; it is also difficult to study surfaces, which are usually uneven, and it is impossible to study bulk objects such as engineered parts or whole creatures.

The invention which changed this state of affairs was the SEM. In the SEM, the electrons do not pass through the specimen, but are 'reflected' from its surface. The history of the SEM is one with several false starts. The scanning principle was first demonstrated in 1935 by Max Knoll, who had gone to work for Telefunken on the development of television. He obtained a life-size image of a piece of transformer core using electrons to scan its surface. Three years later, Manfred von Ardenne, working in Berlin in association with Siemens, introduced the idea of using an electron lens to produce an ultrafine probe of electrons. This principle is fundamental to all modern SEMs. But von Ardenne was more interested in the TEM, and specimen preparation techniques associated with it. Scanning microscopy seemed relatively unimportant.

It resurfaced in Cambridge in 1948 under the supervision of Charles Oatley, who had an engineer's rather than a biologist's view of the machine. He was interested, in particular, in its ability to examine the surfaces of solid specimens, and in the possibility that contrast in the image might be generated by differences in the atomic structure of the specimen. As with the TEM, development of the SEM was assigned to a Ph.D. student, Dennis McMullan. During the 1950s, it became clear that image contrast could result simply from the specimen's surface shape rather than from its composition, and this method of producing the image is the most widely used today. It means that the microscope can be used with biological specimens, which only need a thin

coating of a metallic conductor (usually gold) in order to produce a high-contrast imaging signal.

The first commercial SEM, the Stereoscan, was marketed in 1965 by the Cambridge Instrument Company. With its development, electron microscopy came of age. Microscopists had two basic and complementary electron instruments with which to study an almost unlimited range of specimens at much higher magnifications than is possible with the light microscope.

The story of modern microscope development considered from the standpoint of the instrument alone omits an enormously important point, namely that it has revolutionised our view of the world. The electron microscope in particular has opened new paths of discovery. Nor does the story end with the microscopes so far described. New microscopes are being developed which exploit X-rays, sound waves and even the electron clouds which surround all matter to form the image. The modern microscopist is far removed indeed from Robert Hooke, who built his own instruments and drew pictures of what he saw. One thing has not changed in three centuries, however. The motive for using microscopes remains what it always was – the delight in being able to see and show to others the invisible world beyond the eye.

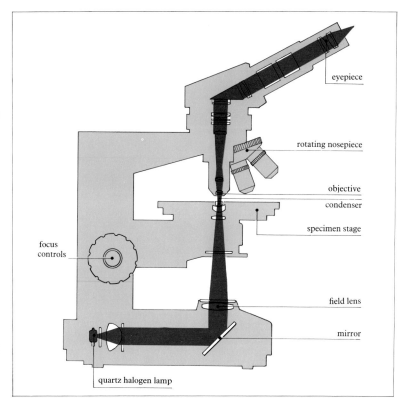

The optical path through a modern light microscope. Light from a high-intensity lamp in the base of the instrument is reflected by a mirror into the bottom of the condenser. It passes through the specimen, enters the objective lens, and finally reaches the eyepiece after having been refracted through inclined prisms in the viewing head. This arrangement gives maximum mechanical stability coupled to high user convenience.

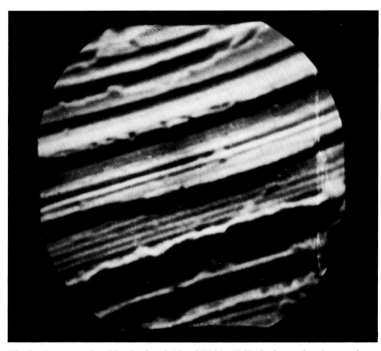

The first image produced by the Cambridge SEM in 1952. It shows abrasions on the surface of a piece of mild steel, at a magnification of ×3000.

LIGHT MICROSCOPY

Light microscopes come in a variety of forms, ranging from instruments which are scarcely more than toys to research tools with a host of facilities. But they all share the same general design and principle of operation. Most light microscopes are designed for viewing transparent specimens in transmitted light, and are used mainly for biological studies. Microscopes used in metallurgy, geology and the materials sciences are treated here as a speciality. The optical principles are the same in both areas, but differences in lighting methods and specimen preparation are described where relevant.

A light microscope consists of a stand with the optical components attached to it. The stand provides a stable frame which holds the lenses, the specimen and the illumination system in a precise relationship to one another. Modern microscopes consist of a rigid stand with a viewing head at the top which is tilted towards the eye for convenience in use.

Attached to the microscope stand is

the tube which holds the image-forming lenses. The tube is of standard dimensions to allow interchange of both objective lenses and eyepieces, although different manufacturers may adopt different tube lengths. The most common are 160 millimetres and 170 millimetres, and it is essential to use objectives marked with the appropriate figure. In research microscopes, light may emerge from the objective lens as a parallel beam – so-called infinity-corrected optics – which means that the distance between objective and eyepiece is no longer of critical importance.

Focusing of the image is achieved by varying the distance between the top of the specimen and the bottom of the objective lens. This movement is controlled by coarse and fine focus knobs which act on very precisely engineered gears, often self-compensating for wear.

The specimen stage is the platform on which the specimen is placed. At its simplest it may be fitted with just two spring clips to hold a glass slide. On research microscopes the stage has calibrated mechanical controls to allow the specimen to be moved precisely in two dimensions. It can be rotated about its centre, and in the case of polarising microscopes this rotation is also precisely calibrated.

Beneath the specimen stage is the substage illumination system. It is here that microscopes differ most. The most basic substage consists of no more than a mirror to reflect light from an external source up through the specimen. Better quality students' microscopes, and all research microscopes, have a condenser system as part of the substage. A condenser consists of an arrangement of lenses which can be centred and focused up and down. It is also fitted with a variable aperture diaphragm. Condensers can provide very bright illumination of the specimen in such a way as to exploit the properties of the objective lens to the full. Good condensers have facilities for phase contrast and interference microscopy. Their optical properties match those of the objective lens, and they are very expensive. The substage also has facilities for the attachment of filters and polarising devices.

The Objective Lens

The optical components of the microscope comprise the two imaging lenses in the tube, and the condenser. Each plays a different role in image formation. The lens at the top of the microscope tube is the eyepiece. The lens at the bottom is the objective, and it is the main determinant of image quality.

The objective is always a highly corrected lens containing several glass elements – a so-called compound lens. Most microscopes have a rotating nosepiece which holds several objectives of different magnification. The magnification of an objective lens is inversely related to its focal length: the shorter the focal length, the higher the magnification.

The magnification of the final image in the microscope is the product of the magnification of the objective and that of the eyepiece. As a result, different magnifications are achieved by rotating the nosepiece to bring a different objective into play. In modern microscopes this is very convenient, since the lens mounts are designed so that changing magnification does not result in loss of focus.

Objectives come in three qualities. The cheapest and simplest are called achromats. They are corrected for

chromatic and spherical aberration at two points in the visible spectrum, in the blue and red regions. They are adequate for visual observation, or black-and-white photography.

The use of the mineral fluorite or fluorite-like glasses in an objective means that a higher degree of optical correction is possible. These lenses form the next quality of objective, and are designated 'fluorite' or 'Neofluor' on the mount. They generally have larger numerical apertures than the corresponding achromats and yield a brighter, more contrasty image with finer detail. They are suitable for colour photography, being corrected at three points in the spectrum for chromatic and spherical aberration.

The best objectives are called apochromats. Modern apochromats by leading manufacturers are corrected at four wavelengths in the spectrum for chromatic and spherical aberration. They have large numerical apertures and provide the best possible image in terms of resolution and freedom from optical defects. They are also very expensive.

All these lens types suffer from a

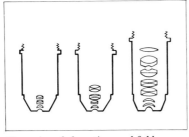

Correction of aberrations and field curvature in objective lenses involves the use of many glass elements of different composition, shape and spacing. The diagram shows the elements in a typical achromat (left), apochromat (centre) and planapochromat (right).

fault known as curvature of field: when a flat specimen is examined, it is not possible to focus sharply the centre and the edges of the image simultaneously. This is of little importance for visual observation, since microscopists continually move the specimen and adjust focus in order to build up a mental picture of the object under study. But it is very undesirable in high-quality photomicrography, and manufacturers therefore supply objectives in the three qualities already mentioned, but with the additional correction of flatness of field. They are designated on the mount with the prefix 'plan' – 'plan-

apochromat', for example.

Apart from its quality designation, an objective lens may have other markings on its mount. 'Ph', for example, means a phase contrast objective; 'Pol' indicates a strain-free objective suitable for high-quality work with polarised light. In addition, all lenses carry two pairs of numbers. These refer to the magnification of the lens, paired with its numerical aperture, and to the correct tube length together with the correct thickness of cover slip (the cover slip is the thin piece of glass placed on top of the specimen). Thus a lens might be marked

40/0.65

170/0.17

This means that it has a magnification of ×40, a numerical aperture of 0.65, and should be used on a microscope with a tube length of 170 millimetres and a cover slip 0.17 millimetres thick. Lenses marked 'Met' (metallurgical) are designed for use without a cover slip.

The purpose of the objective lens is to produce an image with as much fine detail as possible. It must be able to form an image of the space between closely adjacent structures in an object. The minimum distance between two structures which can be distinguished by any particular lens is called its resolution. A high-resolution lens can form an image of very fine detail; a low-resolution lens blurs the finest details and does not produce a really sharp image.

When light passes through a specimen, some of the rays are undeflected, but some are bent – or diffracted – by the specimen. It was Ernst Abbé who first showed that the resolution of an objective lens depends on its ability to collect the diffracted rays of light as well as the undeflected ones. He expressed this ability of a lens to collect light in terms of a concept called numerical aperture.

Numerical aperture is defined by the relationship:

$$\text{numerical aperture} = n\mathrm{Sin}a$$

In this expression, n is the refractive index of the medium between the top of the cover slip and the bottom of the objective lens, and a is half the angle of acceptance of the lens. At its simplest, the equation shows that for a given magnification, the larger the front glass of the lens, the better its resolution (since a is larger).

Numerical aperture is itself directly related to the resolving power of the lens by the equation:

$$\text{resolving power} = \frac{0.61 \times \text{wavelength of light in use}}{\text{numerical aperture}}$$

This equation tells us two things: first, that the larger the numerical aperture of the lens, the higher its resolution; secondly, that the shorter the wavelength of light in use, the higher the resolution within the image.

Increasing the numerical aperture of a lens of given magnification improves the resolution but results in a decrease in the depth of field. This is not usually of practical significance for specimens which are thin sections or highly polished flat surfaces, but it does limit the examination of bulky or uneven objects common in materials science.

Numerical aperture (and hence resolution) depends not only on the angle of acceptance of the objective, but also on the refractive index of the medium between it and the cover slip. For air, the refractive index is 1. This means that the highest numerical aperture possible for a lens used in air is about 0.95 (since the angle of acceptance cannot exceed 180 degrees, and in practice is always less). To improve the resolution of high-power objectives, they are usually designed as 'immersion' lenses. A drop of oil or, less frequently, water or glycerol, is placed between the top of the cover slip and the bottom of the lens. The effect is to increase the numerical aperture by a factor equal to the refractive index of the immersion medium. For most immersion oils, this factor is just over 1.5. The maximum possible numerical aperture of an oil immersion objective is thus about 1.4, and its resolution will be correspondingly better than a comparable dry lens.

Immersion lenses are always marked on the mount. 'Oel' means that immersion oil should be used, 'W.I.' designates water, and 'Glyz.' glycerol. These lenses give poor results if used in air.

The Eyepiece

The function of the eyepiece is to magnify further the image produced by the objective. Eyepieces contain two lenses and a fixed aperture called a field stop. Modern microscopes tend to have so-called compensating eyepieces, which are designed to compensate for small residual defects in the primary image from the objective. Other eyepieces have a high eye point, for use by spectacle wearers, or are designated wide field because they allow examination of a greater area of the primary image in the microscope tube. Many microscopes, of course, are fitted with binocular heads containing two eyepieces.

Eyepieces usually come in a range of magnifications from ×8 to ×25. The final magnification of the perceived image is the product of the magnification of the objective lens and the eyepiece. Thus a ×25 objective and a ×10 eyepiece give a final magnification of ×250. So too does a ×10 objective and ×25 eyepiece. However these images will not be the same, since the numerical aperture of a ×25 objective will almost certainly be higher than that of a ×10 objective; hence the image produced by the former will contain more resolved detail than the image from the latter.

The Illumination System

Early microscopes used daylight or lamp light reflected by mirrors as the source of specimen illumination. In modern microscopes, the light source is invariably an electric lamp. It may be quite separate from the microscope stand, consisting of an opal bulb fitted inside a housing equipped with a variable aperture exit diaphragm. This arrangement is inexpensive and suited to microscopes with just a substage mirror, or a mirror and an Abbé type condenser. But it is unsatisfactory for critical study, and especially for photomicrography.

The best modern microscopes have their own light source built into the base or back of the stand. The lamp is usually a low-voltage quartz halogen type, which gives intensely bright illumination of the correct temperature for colour photography. The light is deflected into the base of the condenser by means of a fixed mirror and through a variable aperture diaphragm.

The substage condenser consists of a compound lens fitted with a variable aperture, and fixed within a focusing mount which can also be centred. The function of the condenser is to illuminate the specimen in such a way that the front glass of the objective lens is filled with light. This means that the requirements for a condenser are

exacting and to some extent contradictory, since it will be used with objectives of different magnifications. When a low-power, ×2.5 objective is in use, the condenser is required to illuminate a circle of the specimen about 7 millimetres in diameter with a bright, even and narrowly converging beam of light. With a ×100 immersion objective, on the other hand, it is required to illuminate a field only 0.2 millimetres in diameter, but with a strongly converging cone of light. A compromise is commonly reached whereby one of the lenses in the condenser can be swung out of the light path when the system is used with particular types of objective.

Methods of Illumination

The typical microscope specimen is transparent, mounted on a glass slide beneath a cover slip and viewed with transmitted light. If the object is too bulky to be transparent, the specimen usually consists of a thin section of it. Some objects have intrinsic colour and good contrast, and can be viewed straight away. Others need to be stained with dyes before structures within them become visible.

For the examination of a stained section on a slide, all that is required is a light source and a properly adjusted condenser. With a separate light source, the condenser is arranged so as to project an image of the surface of the opal bulb into the specimen plane, the condenser itself being centred. This is Nelson or 'critical' *bright field illumination*. With an integral light source, the favoured viewing condition is Köhler illumination. The centred condenser is focused so that it projects an image of the field diaphragm in the base of the microscope into the specimen plane.

In practice both bright field methods require the condenser to be set high in its focusing mount, almost touching the bottom of the slide. This gives maximum resolving power, a bright image and an evenly illuminated field. The brightness is controlled by adjusting the intensity of the lamp, not by altering the height of the condenser.

With bright field illumination, contrast has to be achieved either by the specimen's intrinsic properties or by staining. Other illumination methods, however, enhance image contrast. The most common of them

are dark field illumination, phase contrast, and differential interference contrast.

The principle of *dark field illumination* is that direct rays from the light source are prevented from entering the objective lens. This can be achieved by fitting a patch stop to the condenser to block out the central part of the light beam. Also available are dark field condensers of sophisticated design.

The benefit of dark field illumination is best seen with

suspensions of small particles such as yeast or bacterial cells, which are difficult to see in bright field conditions unless stained. In dark field, the background is dark and the cells stand out clearly as points of intense brightness. Colour contrast can be created by using annular filters in the condenser. For example, a filter with a blue centre and a yellow periphery produces a blue background against which the transparent specimen appears yellow. This is 'Rheinberg illumination' and it can

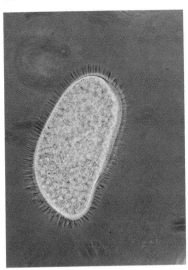
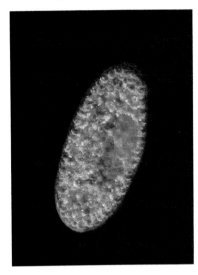
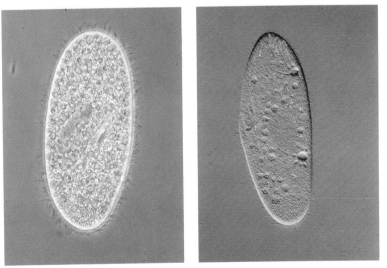

These four pictures are all of the same organism, the ciliate protozoan Paramecium. *In conventional transmitted light (top left), the specimen appears in its natural colours. The green particles are algal cells within the ciliate; the fine bristles at the surface are the cilia. Dark field illumination (top right) enhances the drama of the image and emphasises the presence of refractile particles within the organism; the cilia, however, are not visible. In phase contrast (bottom left), colour saturation is reduced due to flare around the edges of refractile objects; but more detail is visible within the cytoplasm of the organism. By using the Nomarski differential interference contrast technique (bottom right), the image is transformed into a three-dimensional illusion. The algal cells appear as depressions within the ciliate cytoplasm, and resolution of fine detail elsewhere is improved. This is due to the optical sectioning effect of the illumination method. Colour information, however, is almost completely lost.*

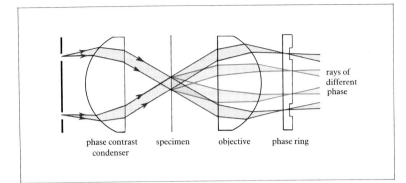

The optical arrangement for phase contrast illumination. The phase contrast condenser produces a hollow cone of light (red). As the light encounters the specimen, some passes through it undeflected (red), and some is diffracted (blue). The undeflected rays encounter the retarding ring in the phase contrast objective, whereas the diffracted rays do not. When the rays recombine above the objective, contrast results from their interference.

produce spectacular results with unstained sections and small aquatic organisms.

In ordinary transmitted light, the image is formed by recombination of undeflected and diffracted rays of light. In *phase contrast* optics, matters are so arranged that this act of recombination produces greatly enhanced contrast. It requires a special condenser and special objectives. Within the condenser there is an annular aperture in the light path, and this produces a hollow cone of light which passes through the specimen. Some of this light travels undeflected and enters the objective, where it encounters a so-called phase ring which has the effect of retarding the light by a distance corresponding to one quarter of a wavelength of green light. Meanwhile, light that was diffracted during its passage through the specimen enters the objective so as to pass on either side of the phase ring.

Thus the light leaving the objective lens contains two components which are slightly out of phase with each other.

The contrast enhancement depends upon a further factor. The undeflected rays are retarded not only by the phase ring, but also by their passage through the specimen itself. This retardation is small (up to about a quarter of a wavelength) and variable, but when it is added to the fixed retardation caused by the phase ring, destructive interference occurs between the retarded beam and the unretarded beam of diffracted light. The retardation of one quarter wavelength by the phase ring and up to one quarter wavelength by the specimen means that in some regions of the image the combined retardation is one half of a wavelength. Combining two waves which are out of phase by half a wavelength results in their total extinction. Where this happens in the

specimen, the image is black; and elsewhere it becomes shades of grey, depending on the precise amount of retardation. Because the increase in contrast results from local variations in the refractive index of the specimen and not from artificial treatment such as staining, phase contrast optics enable the microscopist to produce a high-contrast image of undisturbed living cells.

Phase contrast suffers from one important defect: the image is always accompanied by bright haloes of light around sharply delineated edges of refractive parts of the specimen. This defect can be overcome by using a more complicated interference system based on polarised light. Many such *differential interference contrast* (DIC) systems have been designed, and they all differ slightly. The description which follows is of the Nomarski DIC microscope.

Light from the microscope lamp is first polarised by passing through a polarising filter. It then enters a prism which splits it into two beams and also rotates the plane of polarisation of each so that they are polarised at right angles to one another. The beams pass through the specimen, separated by a very small distance. They then pass through the objective lens, recombine and finally interfere with each other after passing through a second polarising filter below the eyepiece. The effect of this extremely complicated optical arrangement is that the viewer sees a high-resolution image of a very shallow 'optical section' of the specimen. The Nomarski system reduces the visual effect of out-of-focus parts of the specimen above and below the actual focal plane, and this produces an unusually clear image which also has a strikingly three-dimensional appearance – an illusion caused by the formation of shadows on one side of particulate structures within the specimen.

Many specimens are unsuitable for examination by transmitted light. They may be opaque, such as metals and minerals, and although it is possible to make thin sections of such objects, this is not always desirable. In many industrial applications, for example in the textile or microchip industries, the specimen has to be examined intact. In such cases, *incident light* methods must be used.

Incident lighting is obtained most simply by directing the output from a

free-standing lamp onto the surface of a specimen. This is a crude method which can be used only with lenses which work at a sufficient distance from the specimen. Its most common application is with the stereo microscope. Incident lighting can also be obtained by using a Lieberkühn attachment, a concave mirror fitted around the objective lens which directs light coming from below back onto the upper surface of the specimen.

Modern practice uses a more refined incident technique. The light source is fitted behind the microscope tube, and its output is directed onto a semisilvered mirror located between the objective and eyepiece lenses within the tube. Light from the lamp travels down through the objective lens, which thus acts as its own condenser, and is reflected back from the specimen surface, returning upwards through the objective, through the semisilvered mirror, and into the eyepiece. This arrangement places no restriction on the distance between objective and specimen, and is the standard method of illumination in metallurgical microscopes.

A specialised use of incident lighting is found in the *ultraviolet fluorescence* microscope. The principle underlying fluorescence is that many natural materials and synthetic dyestuffs can absorb light of one wavelength and emit the energy as light of a longer wavelength. Ultraviolet light is invisible to the human eye, but a specimen irradiated by it can give rise to visible contrast by the fluorescence it induces. Incident lighting is not essential for ultraviolet fluorescence, but modern practice favours its use.

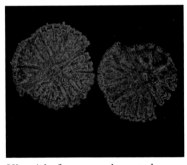

Ultraviolet fluorescence does not always involve the use of artificial dyes as fluorescent stains. Some natural pigments produce intense fluorescence when irradiated with ultraviolet light. This picture is of two desmids, and was taken using a fluorescence microscope. The intense red colouring corresponds to chlorophyll.

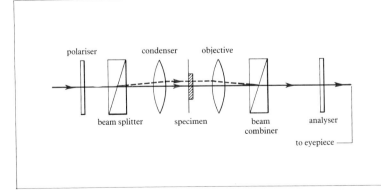

The optical arrangement of the Nomarski differential interference contrast microscope. Polarised light passes through a beam splitter and encounters the specimen as two rays with a slight relative displacement, polarised at right angles to one another. Above the objective the beams are combined and interfere to produce the contrast effect.

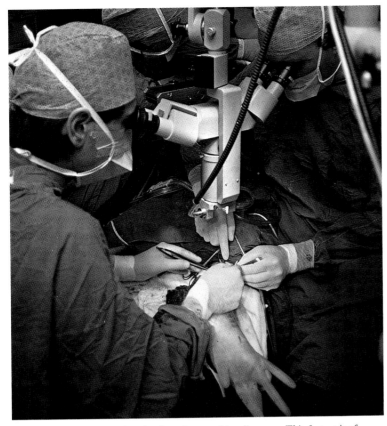

Stereo microscopes are designed to have long working distances. This feature is of particular use in surgery, where unrestricted access beneath the objective lenses is essential. This particular microscope is also fitted with dual binocular viewing heads, enabling two surgeons to view the subject simultaneously.

Light from an ultraviolet source built into the back of the microscope enters the tube between the objective and eyepiece lenses, and is deflected downwards by a special mirror. The ultraviolet light passes through the objective and irradiates the specimen. Fluorescent regions of the specimen emit visible, longer wavelength light as a result, and this is used to form the image. The special properties of the mirror in the microscope tube mean that only visible light can pass up into the eyepiece. The operator thus sees an image consisting of a black background against which any fluorescent structure within the specimen is brilliantly lit.

The ability to produce antibodies to specific proteins within cells has made ultraviolet fluorescence a powerful tool in biology. The antibodies are chemically tagged with fluorescent dye molecules, and are then used to stain a specimen. The bright spots of fluorescence in the image indicate the positions of the antibodies, and hence the proteins for which they are specific. The technique has made possible rapid and specific screening tests for pathogenic organisms, including small bacteria and viruses.

Microscopes for Specific Purposes

The conventional light microscope design and techniques described so far are used throughout science, medicine and industry. But certain applications require specialist microscopes that merit consideration.

The *stereo microscope*, also known as the dissecting microscope, is designed to allow easy access to the specimen so that it can be manipulated during observation. It has two special features. First, the image seen through the two eyepieces is the right way round (whereas in other microscopes, of course, it is reversed) and truly stereoscopic. This is achieved by having either two separate objectives, or a beam-splitting prism immediately behind a single objective. In either case, the images presented to the left and right eye differ slightly, and the brain synthesises a three-dimensional picture of the specimen.

The second feature of the stereo microscope is the very long working distances of the objective(s). There may be several centimetres between the bottom of the lens(es) and the top of the specimen, allowing easy access for dissection, the assembly of microelectronic circuits, and so on. Stereo microscopes are usually equipped with rudimentary facilities for transmitted and incident light illumination. They are limited to a maximum magnification of about ×200. Magnification is often controlled not by altering the objective but by a 'zoom' system in the viewing head.

The *polarising microscope* is designed to allow precise measurements of the properties of minerals, crystals and other birefringent materials such as wood or textile fibres. Polarised light is produced by means of a high-quality polarising filter beneath the condenser. The analyser consists of a second polarising filter, above the objective. The specimen stage can be rotated as usual with a research microscope, but is calibrated to 0.1 degrees to allow precise measurements of angular movements. Above the objective lens there is a slot in the microscope tube to allow insertion of various devices such as quarter wave plates and quartz wedges, which give quantitative measurements of the refractive index and other optical properties of crystals. Above this slot, and below the eyepiece, is a removable Bertrand lens. This is a converging lens which projects an image of the back focal plane of the objective into the eyepiece. This image, called the interference image, can be used in the identification of crystals and minerals. The polarising microscope is a highly specialised and yet versatile instrument which can be used both with transmitted light for studying crystalline materials, and with incident lighting for examination of metals and alloys.

Interference effects can be used not only to increase contrast, as in DIC optics, but also in quantitative ways. For this purpose a specially designed *interference microscope* is used. The principle of its operation is similar to that of phase contrast and the Nomarski system, in that two beams of light travel slightly different paths and recombine to form the image. In both these techniques, however, it is difficult to extract quantitative information from the image because its appearance depends on local differences within the specimen itself.

In the interference microscope, the beam which has passed through the specimen is compared to a reference beam which has *not* passed through the specimen. Because of this, the image contains information about the specimen on an absolute basis, and can be analysed quantitatively to yield information such as the dry weight of biological materials, the refractive index of minerals or crystals, the thickness of thin films, or the depth of wear and accuracy of contour of machined parts.

Specimen Preparation Techniques

The range of microscope specimens is vast, and the methods of preparing them for examination equally so. The broadest division is into transparent specimens and those which are opaque. This corresponds approximately, but not exactly, to biological specimens on the one hand and inorganic materials on the other.

Some objects are intrinsically transparent and require no special treatment beyond mounting in a suitable medium between a glass slide and cover slip. Biological examples include individual cells and small aquatic organisms. Another wide range of specimens consists of fine particles – bacteria, pollen grains, dust and smoke particles – and these too can be examined in liquid suspension without prior treatment.

Keeping preparation simple is particularly desirable for routine and repetitive work. An example is the making of a blood smear. A droplet of blood is placed on a glass slide and smeared across it using a second piece of glass. After drying for a few moments, such a specimen can be examined to diagnose various abnormal conditions, including the results of infection. The information yielded by such minimal techniques is often limited, but nonetheless adequate for many purposes.

Rapid staining methods can sometimes be used on dried whole specimens such as blood smears or dried films of bacterial cells. An example is the Gram staining procedure for bacteria. The cells are smeared and then fixed to the slide by drying. The preparation is covered for about 30 seconds in a solution of the dye methyl violet. This is poured off and replaced with a solution of iodine in potassium iodide. After another minute or so, the slide is thoroughly

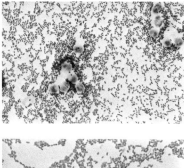

This pair of pictures illustrates the Gram staining technique. The Gram-positive bacteria Streptococcus pneumoniae *have retained the methyl violet and appear purple (top). The Gram-negative* Escherichia coli *(bottom) appear pink.*

washed in alcohol, then stained with neutral red. The result depends on the class of bacteria present on the slide. 'Gram-positive' bacteria, which are generally sensitive to penicillin, retain the methyl violet dye and appear purple. 'Gram-negative' bacteria retain only the second stain, and appear pink or red. Coupled with the shape of the bacterial cells, this simple staining method is a remarkable aid to rapid identification of bacteria and diagnosis of disease.

The great majority of objects, however, are too bulky to be examined by such techniques. The microscopist has to resort to frequently lengthy procedures in order to produce a thin section which can act as the specimen. In biology, this usually means that the whole object has to go through a stage of chemical fixation, followed by substitution of the water within it by an organic solvent, which is in turn replaced by paraffin wax. The object embedded in the wax is then sectioned on a microtome. Before examination, the wax is dissolved away and the section stained.

Staining has been practised for over a century, and is a well-developed art. Stains are available for the major classes of molecules, such as proteins, nucleic acids and other polymers. Two or more are often used in combination to give colour contrast within the specimen. A well-known method from botanical microscopy is to prepare a paraffin wax section, attach it to a

slide, and remove the wax with xylene. The specimen is then washed and stained in a solution of safranin, a red dye, for about an hour. The safranin is then washed away and the slide immersed in a solution of fast green for a few minutes. Next, the slide is passed through two changes of a mixture of clove oil and alcohol, which selectively removes the green dye. The final specimen has high colour contrast: cell nuclei, chromosomes and lignin stain red, while cell walls and cytoplasm stain green.

Colour deposition can also be achieved by means of chemical reactions resulting from the actions of enzymes within the tissue of the specimen itself. This can be used to localise the position of enzymes within cells. Alternatively, antibodies tagged with fluorescent dyes can be used to localise the position of their corresponding antigens.

Staining techniques can also be quantitative. This is exemplified by the Feulgen stain for DNA, which is used to measure the amount of DNA within a single cell nucleus. Further information can be gained by coupling the technique of autoradiography to microscopy. The living tissue is exposed to a chemical, usually a nutrient, the atoms of which have been made radioactive. For instance, some of the ordinary carbon atoms in the nutrient may have been replaced by the radioactive isotope carbon-14. The tissue is allowed to absorb the labelled nutrient and, after an interval of time, is fixed, embedded and sectioned as usual. Next, a thin film of photographic emulsion is laid over the section on the microscope slide, and this sandwich is left for a period of a few days. The radioactive decay of the carbon-14 causes localised exposure of the emulsion at those points where the isotope has concentrated in the tissue. The photographic emulsion is then developed *in situ* and the section stained to complete the preparation of the specimen sandwich. The microscope image consists of a standard view of the tissue, but with the additional information provided by black grains of silver from the developed emulsion, which mark the sites where the carbon-14 was incorporated. This technique gives very precise information about sites of synthesis of biological polymers, storage molecules, cell wall materials and so on. It can also be used to study mechanisms of reproduction of

pathogenic organisms such as viruses.

Non-biological objects are also either transparent or opaque. Examples of the former include crystals of inorganic and organic chemicals, some minerals and a range of finely divided geological materials such as sand. Microscopy is used to study various properties of such materials. The polarising microscope, for example, when equipped with a universal stage capable of orientation in three dimensions, can be used to identify small fragments of crystalline materials from their optical properties. Microscopes can also identify chemicals including tiny amounts of drugs, by examining the form of crystalline derivatives. Such identification may be assisted by the observation of melting points in a microscope fitted with a heating stage.

Other objects only become transparent when considerably thinned. In preparing a mineral specimen, for example, the first stage consists in cutting a section about 1–2 millimetres thick. This is smoothed, but not polished, on one side, and attached by this smooth surface to a microscope slide. The upper face of the section is then ground down with successively finer abrasives until the section is 0.03 millimetres thick. If the specimen is for transmitted light study only, it is now mounted under a cover slip. Surface irregularities will not influence the image if a cement or mountant of appropriate refractive index is selected. Alternatively, if the surface of the specimen is highly polished, both transmitted and incident light techniques can be used. This procedure is broadly applicable to a wide range of objects of industrial interest: ceramics, slags, catalysts. Poorly consolidated materials can be prepared in the same way after they have been embedded in a resin.

If a specimen is opaque, it must be examined by incident light. This may involve nothing more than placing the entire object on the stage of a stereo microscope. An opaque specimen is often an object which cannot be subjected to any preparation by its very nature – a microelectronic circuit, for example, or a bullet fired from an unidentified gun. Many forensic specimens fall into this category, and require a technique called comparative microscopy. This uses a microscope with paired optics arranged so that two specimens are seen side by side. Such

an instrument can be used to determine whether two bullets were fired from the same weapon by comparing the unique pattern of abrasive marks produced by imperfections in a gun's barrel.

Information concerning the internal structure of an opaque specimen can only be gained after careful preparation. Typical is that used for a metallic specimen. The selected specimen is carefully cut to appropriate size, often by high-speed abrasive wheels drenched in cooling water, which do not alter the specimen's microstructure. It is then cleaned and, if it is irregular in shape, it is mounted in plastic, epoxy resin, or bakelite so that it can be bolted into the equipment used to complete the specimen preparation.

The essential requirement is that the surface of the specimen is extremely flat and highly polished. This finish, which must be achieved without damaging the microstructure, is created by grinding with silicon carbide abrasive papers of successively finer grade. Five or six grinding stages may be used, followed by polishing with successively finer diamond grits loaded onto laps covered with cloth. It is usual to start with 7-micrometre diamonds and progress to 1- or ¼-micrometre diamonds in three or four stages. The properties of the material under study will require particular times, pressures, lubricants, abrasives and cloths. In the case of most materials, there are recognised approaches to achieving the necessary finish.

Normal practice is to examine the specimen in its polished state, and then to etch it in order to reveal further details of its microstructure. Etchants are usually liquids containing chemical mixes designed to highlight features in the specimen's microstructure. Specimens such as ceramics may be etched by being heated in air or other gases. For metal samples, selective chemical colouring is frequently employed.

The technique described here can be applied, with slight variations, to all metals, ceramics, rocks, minerals, coals, concretes, glasses and fibres. Composite materials are frequently examined, for example in the electronics industry. When very hard and very soft materials have to be polished simultaneously, the skills of the preparation technician are tested to the full.

ELECTRON MICROSCOPY

Electron microscopes are large, expensive instruments which require careful installation and constant maintenance. They are almost never encountered outside the laboratory. Particular instruments vary from model to model and manufacturer to manufacturer. This section therefore concentrates on their general design and broad applications. To become a proficient electron microscopist is a process of years rather than weeks or months. This is partly due to the complexity of the instrument, but it also reflects the fact that the electron microscopist has considerably more control over the quality of the final image than does the light microscopist.

General Design

Electron microscopes are of two basic types – transmission electron microscopes (TEMs) and scanning electron microscopes (SEMs). The difference between them lies in the way the electron beam is controlled to form the final image. To begin with, however, we examine the general features which all electron microscopes have in common.

The microscope consists of three separate but interacting parts. First, there is a system of vacuum pumps which removes the air from the space inside the microscope tube or 'column'. A good vacuum is necessary to allow the electrons to move unimpeded down the column (which may be a metre in length) and to minimise contamination of the specimen resulting from interactions between the electron beam and residual gas molecules.

The second part of the electron microscope is the power supply. The lenses in an electron microscope are electromagnetic – they consist of a coil of wire through which a current flows, producing a magnetic field. Functions such as brightness, focus and magnification are achieved by altering the levels of electric current in the various lenses. The control of these currents must be very precise, and they must remain stable once set. This is one function of the power supply. Another is to provide the high voltage necessary to accelerate the electrons towards the specimen. In a conventional TEM this requires a stable voltage of 40 000–100 000 volts. In the SEM, the voltage can usually be

varied between about 500 and 40 000 volts. The voltage determines the wavelength of the electrons in the beam, and it must be held constant to about 1 part in 100 000 or better if chromatic aberration is to be acceptably low.

The third part of the microscope is the column itself. This consists of a series of electromagnetic lenses stacked on top of one another. The coil of wire that forms each lens is shielded from external magnetic fields and cooled by refrigerated water. At the top of the column is the chamber housing the source of the electrons, the 'electron gun'. This is a heated tungsten wire with its associated electrodes.

The specimen chamber in a TEM is about half way down the column. In an SEM it is at the base of the column. The specimen is always inserted through an air lock to prevent loss of vacuum in the column.

Image recording is quite different in the two classes of microscope. In the TEM, sensitive material (film or photographic plates) is introduced into the base of the column and directly exposed to the electrons in the vacuum. In the SEM, the visible image is produced on a cathode ray tube screen which may be remote from the microscope itself, and it is recorded by conventional photography using a camera focused on the screen.

Optical lenses work because rays of light are bent as they pass between media of different refractive index. Lenses for light microscopy consist of many pieces of glass differing both in refractive index and shape; by altering these two variables, the lens designer is able to overcome aberrations and produce a lens of chosen

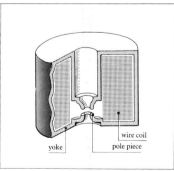

An electron lens consists of a coil of wire surrounded by an iron yoke with a pole piece to concentrate the magnetic field. In practice, such lenses are cooled by water circulating in a jacket which surrounds the yoke.

characteristics. Electron lenses do not work in this way. As the electrons pass through an electron lens, they do not encounter physical inhomogeneities corresponding to air-to-glass interfaces. They travel through a vacuum. However, they do encounter, and are influenced by, the magnetic field within the core of the lens.

Several important design features result from the use of electromagnetic fields to refract the electron beam. The most obvious is that the focal length of the lens is related to the strength of the magnetic field, and hence to the size of the current flowing in the coil of the lens. This means that functions such as focus and magnification are controlled simply by altering electric currents; they do not require physical movements or the exchange of objectives, as they do in the case of a light microscope.

Another consequence is that electron lenses can only be made as positive elements – converging lenses, in optical terms. This imposes a great restraint on the performance of the microscope, since aberrations cannot be corrected by the manufacture of 'compound' lenses. In practice this means that electron lenses always operate at very small apertures; as a result, the best resolution of an electron microscope, although a vast improvement on the light microscope, is very limited compared to the potential resolution which could be achieved with high-aperture lenses working at the same wavelength.

Image Formation in the TEM

The illumination system of the TEM consists of the electron gun and one or, more usually, two condenser lenses. Electrons leaving the heated tungsten filament that forms the electron gun are accelerated towards an anode plate. They pass a biased grid (the Wehnelt cylinder) and travel through a hole in the anode plate before beginning their journey down the microscope column. First they encounter the condenser lenses, which concentrate the electron beam and bring it to a point of focus some way above the specimen plane. The condenser system contains a small physical aperture in the form of a disc of metal such as platinum or molybdenum with a precisely circular hole at its centre. This aperture controls both the intensity and the angle of convergence of the electron beam. The operator usually has a choice of three aperture sizes, from a

hole 25 micrometres in diameter up to 100 micrometres. Aperture size, together with the precise level of focus of the condenser lenses, form the operator's control of 'brightness' in the final image.

When the electrons encounter the specimen, one of three things can happen. They may pass through it unimpeded. They may be scattered without loss of energy (elastic scattering). Or they may be inelastically scattered; this involves an exchange of energy between the electron beam and the specimen, and may cause the emission from the specimen of secondary electrons or X-rays. Image contrast in the TEM depends on preventing the inelastically scattered electrons from contributing to the image. Their exclusion is accomplished by a second small aperture just below the specimen. This is the objective aperture and, again, the operator usually has a choice of three aperture sizes, which comprise the microscope's contrast control. (In practice, the major influence on image contrast is specimen preparation, as will be seen later).

The specimen is surrounded by the objective lens (not to be confused with the objective aperture), which magnifies the image of the specimen to a small degree, of the order of ×50. Changes to the current flowing through the objective lens constitute the microscope's 'focus' control. The magnified image from the objective lens is further enlarged by two other lenses below the specimen. They are called the intermediate lens and the projector lens. Both the absolute and relative excitations of these lenses are controlled electronically by a simple 'magnification' control on the microscope's operating console. Most modern TEMs can be set to give a final magnification anywhere in the range ×1000–×500 000.

Finally, electrons leave the projector lens and strike a screen coated with fluorescent material. The operator sees a continuous image of the specimen, usually green in colour, through a viewing window. Fine adjustments to the image are made while viewing it with an externally mounted, low-power binocular microscope.

The control which the operator has over the image is considerable. Brightness, focus, magnification and to some extent contrast are all adjustable. In addition, the operator can improve image quality by careful

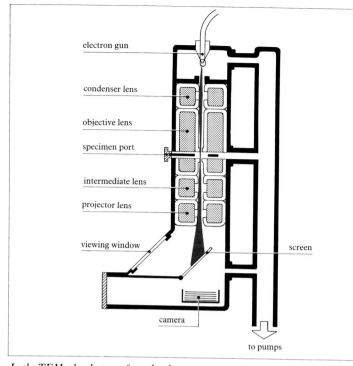

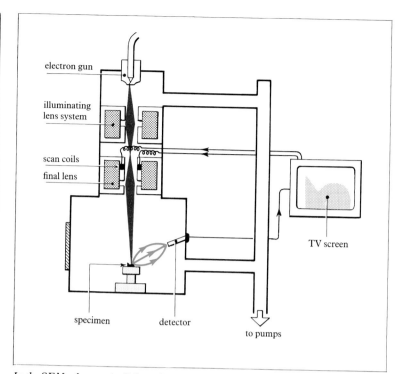

In the TEM, the electrons from the electron gun pass through a condenser lens before encountering the specimen, within the objective lens. Most of the magnification is accomplished by the intermediate and projector lenses. The image is viewed through a window at the base of the column, and photographed by raising a hinged screen.

In the SEM, electrons (red) from the electron gun are focused to a fine point at the specimen surface by means of the lens system. This point is scanned across the specimen under the control of currents in the scan coils situated within the final lens. Secondary electrons (blue) are emitted from the specimen surface and are attracted to the detector. The detector relays signals to an electronic console, and the image appears on a television screen.

adjustments of astigmatism correction controls. However, most of the characteristics of the image are predetermined by techniques used to prepare the specimen. A good operator can get the best out of a specimen, but it needs a well-prepared specimen to produce an excellent image.

Image Formation in the SEM

In the SEM, the image is formed and presented to the operator in a completely different way. The column of an SEM contains an electron gun and electromagnetic lenses corresponding to the condenser system in the TEM. But these lenses are operated in such a way as to produce a *very fine* electron beam, which is focused on the surface of the specimen. The beam is scanned over the specimen in a series of lines and frames called a raster, just like the (much weaker) electron beam in an ordinary television. The raster movement is accomplished by means of small coils of wire carrying the controlling current (the scan coils).

At any given moment, the specimen is bombarded with electrons over a very small area. Several things may happen to these electrons. They may

be elastically reflected from the specimen, with no loss of energy. They may be absorbed by the specimen and give rise to secondary electrons of very low energy, together with X-rays. They may be absorbed and give rise to the emission of visible light (an effect known as cathodoluminescence). And they may give rise to electric currents within the specimen. All these effects can be used to produce an image. By far the most common, however, is image formation by means of the low-energy secondary electrons.

The secondary electrons are selectively attracted to a grid held at a low (50 volt) positive potential with respect to the specimen. Behind the grid is a disc held at about 10 kilovolts positive with respect to the specimen. The disc consists of a layer of scintillant coated with a thin layer of aluminium. The secondary electrons pass through the grid and strike the disc, causing the emission of light from the scintillant. The light is led down a light pipe to a photomultiplier tube, which converts the photons of light into a voltage. The strength of this voltage depends on the number of secondary electrons that are striking the disc. Thus the secondary electrons

produced from a small area of the specimen give rise to a voltage signal of a particular strength. The voltage is led out of the microscope column to an electronic console, where it is processed and amplified to generate a point of brightness on a cathode ray tube (or television) screen. An image is built up simply by scanning the electron beam across the specimen in exact synchrony with the scan of the electron beam in the cathode ray tube.

The SEM does not contain objective, intermediate and projector lenses to magnify the image. Instead, magnification results from the ratio of the area scanned on the specimen to the area of the television screen. Increasing the magnification in an SEM is therefore achieved quite simply by scanning the electron beam over a smaller area of the specimen.

This description of image formation in the SEM is equally applicable to elastically scattered electrons, X-rays, or photons of visible light – except that the detection systems are different in each case. Secondary electron imaging is the most common because it can be used with almost any specimen.

A modern trend in electron microscopy is to fit X-ray analysis

equipment as a bolt-on accessory. Bombarding a specimen with electrons causes X-rays of characteristic wavelengths and energies to be emitted from the spot where the beam strikes the specimen. Computer analysis of the wavelength or energy spectra makes it possible to measure accurately the nature and quantity of different elements in the material. The technique is of little use to biologists because light elements such as carbon produce too weak an X-ray signal. But it is of great value in materials science, particularly because an area as small as 1 square micrometre can be analysed with precision.

Since the output from the SEM is a train of voltages, the operator can exert considerable control over the character of the image. Focus, magnification, brightness and contrast can all be controlled just by turning knobs on the console. In addition, the specimen can usually be tilted and rotated so that it can be examined from a wide range of viewpoints. The output from the microscope can be computer processed so that successive frames are combined and averaged, producing a striking reduction in random noise levels.

Biological Specimens

Electron microscopy places many restraints on the specimen to be examined. First, it has to withstand a high vacuum. This precludes the examination of living materials at ordinary temperatures without prior treatment, except in very special circumstances. In the case of the TEM, the specimen must also be thin. If it is an organic material, it should not exceed 100 nanometres in thickness if good resolution is required. Mineral and metal specimens containing elements of high atomic number must be even thinner because heavy atoms absorb the electrons more readily.

SEM specimens do not need to be thin, but they have other requirements. They must be capable of dissipating the energy of the focused electron beam without building up a local surface electric charge. They must also be a good source of secondary electrons if that is the imaging mode to be used. In practice this last requirement means that the specimen is usually coated with a thin layer of a conducting metal such as gold.

Finally, there is a size limitation on specimens in both types of instrument. This is not very stringent for SEM specimens, but TEM specimens are almost invariably mounted on a fine mesh grid which is 3 millimetres in diameter. This is because of space restraints around the TEM's objective lens.

Biological specimens for the TEM range from single molecules (proteins, nucleic acids, polysaccharides, antibodies) through micro-organisms (viruses, bacteria) to tiny sections of much larger creatures (animals, plants). In general, specimens up to the size of whole bacteria can be studied with no prior treatment other than a staining procedure to enhance contrast.

The most rapid and simple staining technique is 'negative staining'. A suspension of the specimen is mixed with a solution of a heavy metal salt on a specimen grid which has been previously coated with a thin film of carbon. After removing the excess liquid by blotting, the specimen can be examined. What the operator sees is a uniform grey background – the heavy metal salt – with the bright, unstained specimen embedded in it. Negatively stained specimens are naturally very thin, and the technique therefore gives

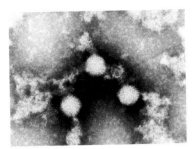

In the negative staining technique, contrast is achieved by the stain forming a grey background, against which the specimen, which is unstained, appears pale grey or white. The stain is most commonly an aqueous solution of a complex salt of a metal such as uranium or tungsten. This picture shows three adenoviruses, a common virus causing sore throats and also implicated in some aspects of cancer.

very high-resolution images. It has been used to elucidate the structure of viruses. It is also very quick – the specimen is ready literally within seconds. This makes it particularly useful when a large number of specimens needs to be examined. In plant pathology, for example, the leaves of a large number of plants may need to be checked for a virus infection; it is a simple matter to negatively stain a drop of sap from each and then examine these specimens for virus particles.

In order to study the structure of cells, and how they combine into plant and animal tissues, other methods have to be adopted. The most common of these involves several stages. First, a small piece of tissue is 'fixed' in an organic aldehyde solution, usually glutaraldehyde. This kills the cells rapidly and stabilises their chemical structure. The fixed material is then placed in a solution of osmic acid, which completes the fixing process and stabilises lipids within the cells, as well as adding electron contrast in the form of metallic osmium. The water in the specimen is next removed, a process called dehydration, and replaced by alcohol. The piece of tissue is then embedded in a liquid plastic such as Araldite. After curing in an oven, the tissue is at last fully stabilised and set in a hard plastic matrix. Slices of the embedded tissue can now be made with an ultramicrotome, placed on specimen grids, and subjected to staining procedures in order to enhance image contrast.

For simple examinations of structure, staining usually consists of

immersing the specimen in solutions of uranium salts or lead salts, or both. Alternatively, specialised stains may be used. For example, if a specimen is floated on a solution containing a hydrolysable phosphate compound and a soluble lead salt, hydrolytic enzymes within the specimen will reveal their position through deposits of insoluble lead phosphate. Proteins which are not enzymes, and other antigenic molecules, can be localised by the ause of antibodies tagged with colloidal gold particles. In both cases, the image consists of a view of the structure of the tissue overlaid with localised deposits of heavy metal (lead phosphate, gold) which show the precise position of a particular molecule under investigation. This is a very powerful technique in biology.

Electron microscopy is fraught with the possibility of obtaining false images, and its practitioners are often accused of seeing what they want to see. Whoever sees a face here is mistakenly identifying some poorly fixed cell membranes.

There is always doubt whether structures seen after chemical fixation are an accurate reflection of the corresponding structure *in vivo*. This has led to methods of preparing biological specimens by freezing. In so-called freeze substitution, the object is rapidly frozen in liquid nitrogen or liquid propane (at about −200 degrees centigrade). This results in the water within it becoming a solid without the formation of ice crystals – it is 'vitrified'. The vitrified water can then be replaced, at low temperature, by an organic solvent such as acetone. The dehydrated specimen is then brought up to room temperature and processed normally.

In cryopreservation, the object is

frozen as before, but it is then sectioned at low temperature with a special microtome. The advantage of this is that the ability of antigens within the tissue to bind to applied antibodies is not impaired. Its disadvantage is that structural preservation may be poor, even if the material is chemically fixed after sectioning and staining.

Another major technique for preparing biological specimens is replica formation, and it is also applicable to a range of materials in metallurgy, engineering and mineralogy. The specimen which goes into the microscope is neither the whole object (as in negative staining) nor a section of it, but a thin film of carbon which has been evaporated onto the specimen in a vacuum. Two applications of replica techniques are worth detailed consideration.

The replica may be of an *external surface* of an object, for example a group of virus particles. The particles are dried onto a suitable substrate, such as clean glass or freshly cleaved mica. This preparation is placed inside a vacuum evaporator and coated with a film of carbon. The carbon forms a coherent layer over the substrate and the virus particles, and the specimen is then produced by floating the carbon layer off the substrate and dissolving away the virus particles underneath. Such a replica, because of its thinness, will yield a very high-resolution image of the surface topography of the replicated particles. Its value as a specimen can be increased by evaporating a layer of a heavy metal such as platinum onto it at a low angle

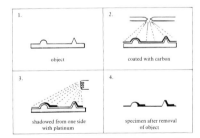

Steps in making a shadowed replica. The object (1) is coated with an even layer of carbon (2). This forms a continuous base for the deposition on one side of a heavy metal such as platinum (3). The specimen is produced by dissolving the original object away, using strong acids, concentrated alkali, or bleach (4). Knowing the geometry of stage 3, it is possible to calculate the height of features on the object from the length of shadows seen in the specimen.

of incidence. This strengthens the replica film; more important, it increases contrast by the production of 'shadows' from highpoints in the replica.

Alternatively, replicas may be made of *internal surfaces* of cells and tissues. This procedure requires much more elaborate equipment, and is called freeze-fracturing or freeze-etching, depending on what happens to the object before it is replicated. The object is first frozen rapidly by plunging it into a mixture of liquid and solid nitrogen, or by propelling it onto the surface of a copper disc cooled by liquid helium ('freeze-slamming'), so that the water within the superficial layers of the object is vitrified. The frozen object is then transferred to the main apparatus, which consists of a vacuum evaporator fitted with sophisticated temperature controls, a microtome and electrodes for the evaporation of carbon and heavy metals. Once inside the machine, the

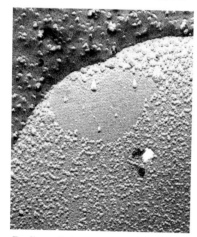

Replica techniques combine high resolution with the ability to look at quite large areas of surfaces. They are particularly useful for studying the surfaces of membranes. In this picture of the outer surface of a photosynthetic membrane from a chloroplast, a patch of regularly spaced particles is visible. The picture could not have been obtained by any other preparation technique. The magnification is ×50 000.

object is sliced at low temperature to reveal internal surfaces. A shadowed replica may then be taken of these surfaces by the evaporation of a thin carbon film followed by a coating of platinum. The microscope specimen is produced by removing the object from the machine and dissolving away organic material with strong acids.

Freeze-etching introduces the additional stage of raising the temperature of the sliced and frozen object to about −100 degrees centigrade for a short period. This causes water within the surface of the object to sublime away, leaving non-volatile components set in sharper relief. Replication and shadowing then follow as before.

These techniques are particularly valuable in the study of biological membranes. Frozen membranes fracture down the middle of their two-layered structure. Freeze-fracture techniques thus make it possible to see the 'internal' surfaces of these membranes. This has been of great value in understanding the functioning of membranes such as those involved in photosynthesis.

It should be emphasised that no one method of specimen preparation can be considered 'correct' for a particular object. Different techniques yield different information about the same structure. For example, a thin section of a mitochondrion within a cell will show its position, outline shape and the way in which the membranes are folded; a freeze-fracture replica will show the distribution of particles within and between those membranes; and a negatively stained preparation of isolated membranes will show details of the particles associated with the surface of the membranes.

With the exception of negative staining, most techniques are time-consuming and require expensive ancillary apparatus. It may take several days to produce a viewable specimen from fresh tissue. Fortunately, both replicas and sections are more or less permanent, and can be retained for examination at a later date.

Turning to biological specimens for the SEM, it used to be necessary first to fix and dehydrate them, and then dry them either in air or by the use of liquid carbon dioxide ('critical point drying'). The only exceptions to this were objects robust enough to withstand a vacuum without such treatment – pollen grains, for example, and insects with hard exoskeletons.

The favoured technique for modern SEM work is cryopreservation. The object is rapidly frozen by being plunged into a mixture of liquid and solid nitrogen at about −200 degrees centigrade. This results in the almost instantaneous freezing of the surface of the object and prevents its distortion. The object is then transferred directly to a special specimen stage, inside the

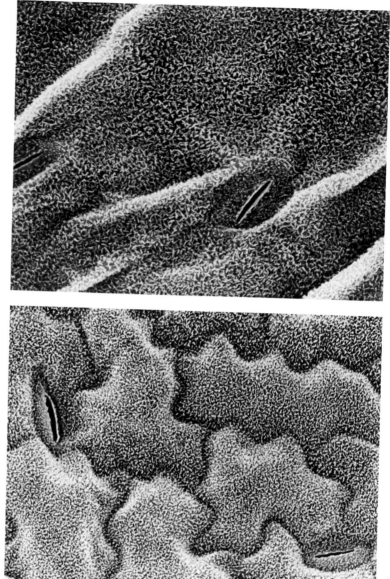

This pair of pictures illustrates the effect of high vacuum on a biological specimen viewed in an SEM. The subject of each picture is the outer surface of a leaf of the garden pea, Pisum sativum. In the top picture, the leaf was placed in the microscope at room temperature after having been coated with a thin layer of gold. The surface appears folded, and although the wax covering the leaf is visible, there is no sign at all of the cellular structure of the epidermis. This is because this piece of the leaf has collapsed and dried under the influence of the high vacuum within the microscope chamber. The lower picture is of the same leaf, but produced in an instrument capable of maintaining the specimen at −190 degrees Centigrade. Again, the wax covering the leaf is visible, but in this picture no collapse or drying out has occurred, and the outline of the epidermal cells of the leaf is clearly seen. The small slits in both pictures are stomatal pores. 'Cryopreservation' of this type is the preferred technique for almost all biological specimens. Both micrographs are at a magnification of ×400.

SEM, which is also cooled by liquid nitrogen. This technique is successful even with the most fragile objects.

In order to achieve good emission of secondary electrons from organic materials and to prevent the build-up of surface electric charge during examination, the specimen is usually coated with a very thin layer of gold or gold-palladium alloy. With dried specimens, this coating can be deposited in a vacuum evaporator. Frozen specimens are usually coated in a special chamber attached to the microscope column and linked to its cooling system.

Cryopreservation does have a few disadvantages compared to conventional chemical fixation and drying. The freezing equipment is costly to buy and run. The specimen stage within the microscope has to be cooled to a very low temperature, and this reduces its mechanical stability. Movements of the specimen stage may also be restricted compared with operation at ambient temperatures. Finally, although the technique is rapid – it typically takes no more than 10 minutes – the resulting specimen is not permanent; once removed from the microscope it is of no further use. The overriding advantage of cryopreservation is that it allows objects to be examined in a fully hydrated state, and without any chemical extraction having taken place.

Materials and Inorganic Specimens

The restraints on the TEM specimen – tolerance to vacuum, transparency and size – apply with slightly different emphasis to non-biological objects. Such specimens include minerals, metals, engineered parts, crystals, dusts and smokes, and a range of biologically derived materials which are nonetheless fairly inert, such as textiles, paper and timber. It is unnecessary to stabilise these specimens against the vacuum of the microscope column. But the problem of lack of transparency may be greatly increased. Biological objects are made from atoms of low atomic number – carbon, oxygen, nitrogen and sulphur. A metal alloy, by contrast, may contain heavy atoms such as lead, tungsten, nickel or uranium. Unless such specimens are very thin indeed, they will scatter the electron beam so much that image intensity will be low and chromatic aberration effects pronounced. As for the limitation on specimen size, it is felt in terms of sampling error. In other words, the small piece of a material selected to produce the specimen may not be truly representative of the bulk.

The preparation of ultrathin specimens can be accomplished in several ways. Many metals can be produced as thin films simply by evaporating them in a vacuum onto the surface of a suitable substrate, which is then dissolved away to leave the thin film specimen. Specimens of metallic compounds can be produced by a similar method, provided that suitable

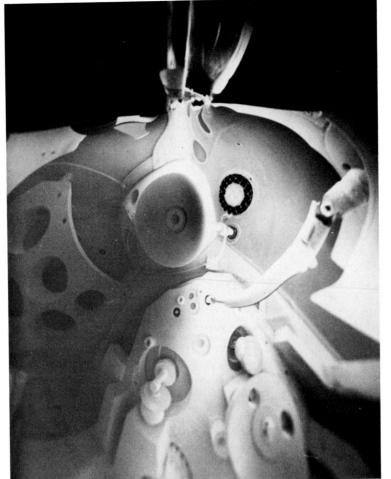

This picture is a self-portrait, taken by an SEM. The effect was achieved by using an insulated specimen, which resulted in primary electrons bouncing from it onto the internal surfaces of the specimen chamber of the microscope. The chamber then emitted secondary electrons, which were detected in the normal way, so that the chamber took its own picture. The view, in effect, is that which would be seen by the specimen on the stage of the microscope. The central round structure with a hole in it is the SEM's final lens; the electron beam emerges from the hole. To its right, caged, is the secondary electron detector. The linkages which are used to move the specimen stage can be seen at the bottom of the picture.

solvents are available. For example, oxides can be studied by the surface oxidation of the parent metal. The layer of oxide is freed from the underlying unreacted material by dissolving the metal in a solvent which does not react with the oxide. Specimens of this type are used to study crystal structure, or mechanical and other properties. The surface relief of materials can also be studied by use of the replica techniques already described for biological specimens.

Of more general application in the production of very thin specimens are methods which begin with a large piece of material and reduce its thickness in stages. Take a rod of alloy steel. The first stage is to produce a slice of it of a size appropriate to the microscope's specimen stage (usually a 3 millimetre diameter cylinder). This can be accomplished by mechanical means (a fine-toothed saw) or chemically by an acid stringsaw. Further stages cannot be carried out mechanically. Methods include chemical thinning, where a fine jet of a dissolving agent is directed onto the centre of the specimen until a hole is produced; regions around the hole are then thin enough to be examined by TEM. Or thinning to the point of puncturing the specimen can be achieved electrolytically, by mounting the specimen as an anode in an electrolytic cell. The edges of the sample can be protected by lacquer during the thinning process, or the

thinning may be localised by directing a stream of electrolyte onto the centre of the sample. Finally, a sample can be thinned by bombarding it with high-energy ions at low pressure. This process involves placing the specimen inside a vacuum chamber, and bleeding in a gas (usually argon) which gives rise to ions when a high-voltage discharge is produced within it. Ion-beam thinning is used with metals, minerals, diamond, carbon fibres and other applications. The thinning can be limited to a particular area of the sample by magnetic focusing of the ions; or the sample can be rotated in the ion beam to give overall thinning on both sides.

With materials such as textile fibres, timber, or asbestos, thin specimens can be produced by sectioning techniques like those described for biological materials. It may only be necessary to embed the sample in a resin suitable for sectioning with diamond knives; chemical fixation and dehydration are superfluous.

A wide range of materials can be placed in the microscope as finely divided fragments. They may be produced mechanically by grinding the original sample (minerals, glasses), or chemically as precipitates. The latter are mounted on specimen grids precoated with a thin film of carbon which acts as a support. In the case of ground fragments, only the edges of individual particles are thin enough for TEM examination.

In many cases, the TEM is inappropriate for examination of materials specimens. Failed machine parts, forensic specimens and manufactured goods undergoing quality control, for example, all need to be studied whole. The nature of the information sought means that destructive methods of specimen preparation cannot be used. In such cases the SEM is the instrument of choice. The elaborate procedures needed to prepare biological SEM samples are largely unnecessary, since stability to a high vacuum and conductivity are often properties of the material itself. Metals, textiles, machine parts and microchips can all be examined without any preparation at all, although in most cases it helps to gold-coat the specimen to prevent local charging effects. Even metallic specimens may need such coating for the best results, since they are almost invariably covered with a thin layer of oxidised material of poor conductivity.

OTHER TYPES OF MICROSCOPY

This section describes some of the specialised types of microscope. Whereas the light microscope, TEM and SEM are widespread throughout the scientific world and indispensable to its work, the microscopes in this section are comparatively rare, due to their expense or specialised function. They demonstrate the way in which not only light and electrons, but sound, X-rays and even the behaviour of atoms can be exploited to produce magnified images of specimens.

Scanning Optical Microscopy

The idea of using a fine beam of light to scan a specimen and produce a magnified image was first put into practice in 1951, in the 'flying spot microscope'. Since then, design and technological developments have resulted in considerable refinement of the scanning optical microscope. The modern instrument is confocal; that is, both the objective and condenser lenses are focused in the same plane as part of the image-forming process. This results in an improvement in resolution by a factor of 1.4. There are two main designs of confocal scanning microscope: one uses laser light; the other, visible or ultraviolet light.

The principle of the laser scan microscope (LSM) is simple. A beam of laser light is focused to a very fine point by means of the objective lens. Focused on the same plane is the condenser lens. Light passes through the condenser, where it is focused onto a pin-hole aperture in front of a photomultiplier tube. Thus at any instant, a point of light from the specimen gives rise to a current in the photomultiplier tube, and this in turn gives rise to a point of light on a television screen. The image is built up by moving the laser beam with respect to the specimen in a raster pattern. Either the light beam itself is moved by means of precisely controlled mirrors, or the specimen is moved on a mechanical stage.

This arrangement gives the instrument two advantages over a conventional light microscope. First, it is confocal, so it has a better resolution. The second is due to the use of points of focused light, and a point source detector (the pin-hole aperture). The image is only formed from light passing through the detector, and this light only comes

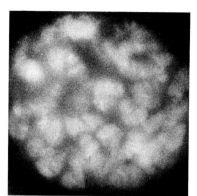

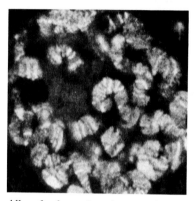

All confocal scanning microscope designs have in common the ability to produce an 'optical section' of an extended translucent specimen. The effect is illustrated in this pair of pictures of the same specimen, a whole nucleus from the salivary gland of a Drosophila *fruit fly larva. Both pictures result from the fluorescence of a dye which has been used to stain the DNA within the chromosomes. In the upper picture, this fluorescence was excited in the conventional way using an ultraviolet fluorescence microscope. It appears blurred because light from above and below the plane of focus of the objective has contributed to the image and degraded its quality. In the lower picture, taken with a confocal microscope, this degradation of the image has not occurred, because only light from the precise plane of focus of the objective and the condenser has contributed to the image. As a result, the outline of the chromosomes themselves can be seen, as well as the pattern of banding along their length. The fluorescence in the lower picture was excited with a laser beam.*

from the precise plane of focus in the specimen. In practice, this means that the microscope has a very shallow depth of field indeed. As a result, up and down focusing movements of the microscope can be used to provide a series of optical sections through the specimen.

Even more remarkable is the effect seen in incident light mode. Here the objective acts as its own condenser, just as in a conventional fluorescence microscope. However, because of the point detector, only those parts of the specimen which are in focus are visible at all. Thus if the microscope is used to look at the tilted surface of a microchip, for example, the image consists of a narrow band corresponding to the region of the specimen which is in focus; the rest is black. If such a tilted specimen is examined while moving the plane of the microscope up and down, the final image shows *every* part of it in focus. In effect the microscope has an indefinite depth of field. This is a useful property, particularly as it is gained with no loss in resolution. The drawback of the LSM is that the image carries no colour information due to the use of monochromatic laser light.

The second type of confocal scanning microscope is called the tandem scanning reflected light microscope (TSRLM). It can use white light, or ultraviolet light. The light passes through a metal plate drilled with thousands of small holes which are arranged in a series of Archimedean spirals. At any one time, light from several hundreds of these holes, which are about 50 micrometres in diameter, passes into the objective lens. The lens focuses the light as a series of several hundred points in a plane within the specimen. Reflected light from this plane then returns through the objective and is focused as a series of points on the underside of the drilled plate. Each of these reflected points is formed at the exact position of holes in the plate. The eyepiece is beyond the plate, and focused upon it. Scanning is accomplished by spinning the plate. This gives the operator a real-time image of the specimen – in colour, if white light is used.

Both microscopes share common advantages in high resolution and the ability to produce a sharp image of planes inside translucent objects such as teeth or bones. The future of such instruments is hard to assess. One potential application is the detection of forgeries, since they can be used in reflected mode to examine layers beneath a coat of varnish, or within an object such as a banknote. A comparable medical use would be the examination of layers under the skin.

High-Voltage Electron Microscopy

The high-voltage electron microscope (HVEM) differs from the conventional TEM in respect of the accelerating voltage which is used to drive the electrons down the column and through the specimen. In the TEM this voltage is up to 100 kilovolts, sufficient for thin sections of biological materials and very thin sections of metallic specimens. The HVEM uses voltages up to 3 million volts.

The principle of operation is the same as in the conventional TEM, but the use of very high voltages results in several practical differences. The first is that the instrument itself is very large and very expensive. Generating stable voltages of the order of 1 million volts requires the use of bulky electronic components. Furthermore, the high energy of the electrons results in a danger to the operator in terms of X-radiation from the microscope, and in consequence the shielding of the column has to be greatly increased to provide protection.

The main operating advantage of the HVEM is that it offers far better penetrating power and resolution when presented with a thick specimen. This was the rationale behind the building of the first, 500 kilovolt HVEM in Manchester in 1952, at a time when neither microtomes nor plastics were sufficiently developed to allow the cutting of thin sections. Nowadays the ability to view thick sections is seen as a positive advantage rather than a solution to a technical difficulty. A thick section of a metal, for example, is more likely to be representative of a bulk sample; and a thick section of a biological object can reveal three-dimensional relationships between cell components. The gain can be considerable; with a 1 million volt machine it is possible to examine biological materials as thick as 1 micrometre – ten to fifteen times the thickness of a typical specimen for a machine operating at 60 kilovolts.

An incidental benefit of the size of the HVEM is that space is much less restricted inside the column, and particularly in the vicinity of the objective lens. This means that a variety of special stages can be fitted. Stages designed to heat or cool the specimen, or subject it to strain, enable the effects of such treatments on new materials to be observed as they happen. The penetrating power of the electrons is such that it is even

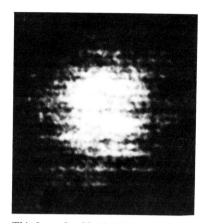

This featureless blur is in fact a single atom of gold. It was produced by a STEM and is printed here at a magnification of ×90 million.

contrast is an electronic function determined by the operator, and resolution, while set by design parameters of the microscope, is not degraded by an imaging lens system. So a microscope which employs the SEM's scanning principle to view the TEM's thin sections should combine the contrast characteristics and lack of lens aberrations of the former with the high resolution due to the latter. This type of microscope is known as the scanning transmission electron microscope, or STEM.

The first STEM was built in 1966 by an American, Albert Crewe. Its essential design feature was the use of a special type of electron gun, called a field emission gun. In a conventional SEM the electron gun consists of a tungsten filament from which electrons are 'boiled off' by heating. The field emission gun consists of a sharply pointed crystal of tungsten in a very strong electric field. The field pulls electrons from the surface of the crystal without the need for heating. The resulting electron beam is extemely intense and, because the crystal acts as a very small diameter source, extremely fine – an essential requirement for good resolution.

The field emission gun only works if the vacuum around the tungsten crystal is very high indeed – of the order of 10^{-10} torr. The cost of this requirement is the main reason why field emission guns are not used routinely in SEMs to give a less noisy and higher resolution image.

The field emission gun's electron beam, focused down to a diameter of just 0.3 nanometres by the microscope's lens system, is moved over the specimen in the same raster pattern as in an SEM. But because the STEM specimen is a thin section, the image-forming electrons are collected after they have passed through it. These transmitted electrons are of two types – elastically scattered and inelastically scattered – and an image can be formed from either. The elastically scattered electrons are by and large deflected through larger angles than the inelastically scattered electrons. Because of this, an annular detector placed beneath the specimen will preferentially collect the elastically scattered electrons. The part of the beam which passes through the centre of the annulus will consist mainly of the inelastically scattered component, and this can be detected separately.

The STEM's output is thus a train of voltages from each of the two detectors. By combining these signals, the image can yield very high contrast and high resolution at the same time. And because the ratio of the signal from the two detectors is dependent upon the atomic number of the elements encountered by the electrons as they pass through the specimen, the instrument can also be used analytically. The STEM requires careful control of operational conditions, and complex analysis of the final outputs. Its use is mostly confined to materials science laboratories.

Acoustic Microscopy

Sound as detected by the human ear has a very long wavelength – middle C, for example, consists of pressure waves with a crest to crest spacing of about 75 centimetres. Such waves are of no use in producing an image in a microscope. However, the wavelength of sound depends on its frequency and also on its velocity (which depends, in turn, upon the medium through which the sound travels). At a frequency of 3 gigahertz (3×10^9 cycles per second), for example, the wavelength of sound travelling in water is only 0.5 micrometres. This is the same as the wavelength of green light, and that fact was first appreciated by the Russian D.Y. Sokolov in 1949. The way was open to design a microscope using sound waves as the imaging illumination, but it was not until 1973 that the first acoustic microscope was built by a team at Stanford University led by Calvin Quate.

The basic component of an acoustic microscope is a lens made of sapphire.

The 1.5 MeV HVEM at the Lawrence Berkeley Laboratory, University of California. Superficially similar in appearance to a conventional TEM, the instrument's column is contained in a three-storey 'silo'; only the bottom part of the column is seen in this photograph.

possible to build stages which allow the examination of specimens at atmospheric pressure, by enclosing the object under study in a cell fitted with thin corundum windows. Using this arrangement, it is possible to study chemical reactions as they occur – between an alloy and its gaseous environment, for example.

Commercial HVEMs commonly run at between 650 kilovolts and 1.2 megavolts. The instrument at Toulouse has an accelerating potential of 3 megavolts. HVEMs are so expensive that they are usually run as a common service by many laboratories within an area. The gradual acceptance of the benefits of higher accelerating voltages, particularly in the materials sciences, has resulted in a generation of TEMs which can operate above the customary 100 kilovolt limit, but do not reach the lower limit of the HVEM range. Their principal application is described in their generic name – high resolution electron microscopes (HREM) – and they are likely to gain widespread use in the coming years. The HVEM however, will probably remain a specialist rarity.

Scanning Transmission Electron Microscopy

Image quality in the conventional TEM depends largely on two conflicting parameters – resolution and contrast. Resolution is maximised by using high accelerating voltages to view very thin specimens. But both the high voltage and the thinness of the section lower the contrast of the image. In the SEM, on the other hand,

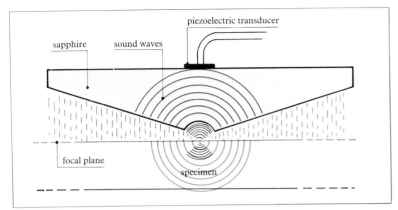

The acoustic lens consists of a piece of sapphire with a spherical depression ground into one face. Opposite this is fixed a piezoelectric transducer, which when excited electrically sets up a mechanical wave in the sapphire. Waves leaving the sapphire through the spherical surface are refracted to produce a fine point of radiation which can be used to scan any chosen plane within the specimen, or at its surface.

It has one flat surface and a spherical depression ground into the opposite surface. A layer of piezoelectric material sandwiched between two gold electrodes is bonded to the flat surface. When a high-frequency voltage is applied across the electrodes, the piezoelectric material (zinc oxide, for example) vibrates mechanically at the same frequency, and sends a sound wave through the sapphire. This is focused to a fine spot as it emerges through the spherical surface on the other side of the lens. The spot, about 0.5 micrometres in diameter, is scanned across the specimen in a raster pattern. This is usually achieved by precisely controlled movements of the specimen itself.

At each point of time during the scan, the specimen reflects the acoustic pulse back into the sapphire, where it travels across the piezoelectric layer and gives rise to a voltage pulse between the gold electrodes. The voltage pulse is used to produce a point of light on a television screen. As in the SEM, image magnification depends on the ratio of the distance scanned to the size of the television screen.

The image is produced by reflection of sound waves, and these, unlike light, can penetrate a visually opaque specimen. If the finely focused spot is set to travel over the specimen surface, then the image is of that surface. But if the spot travels in a plane within the specimen, the image is of that internal plane. This property has given the acoustic microscope an important application in the semiconductor industry, because it makes possible non-destructive examination of the various fabricated layers in a microchip.

Commercial acoustic microscopes use water at room temperature as the coupling medium between the sapphire and the specimen. This produces a resolution comparable to that of a conventional light microscope. Better resolution, down to 0.1 micrometres, has been achieved in research designs by using liquid helium or the gas xenon at high pressure as the coupling medium, but neither of these is likely to result in commercial products.

Photoemission Electron Microscopy

Contrast in any microscope results from the interaction between the specimen and the imaging radiation. Sometimes this interaction is complex; in the SEM, for example, the beam gives rise to secondary electrons, back-scattered electrons and X-rays. A rare event when an electron strikes a specimen is the emission of light – so-called cathodoluminescence. The reverse can also occur – the emission of electrons when light strikes a specimen – and it is this effect which is the basis of the photoemission electron microscope.

The instrument can be compared to an ultraviolet fluorescence microscope. In both cases the specimen is irradiated with an intense beam of ultraviolet light. In the fluorescence microscope, the beam is absorbed and reemitted as visible light. In the photoemission electron microscope, it is reemitted in the form of electrons. The specimen is held at a large negative potential, which drives the electrons through a series of electromagnetic lenses, finally producing an image on a fluorescent screen in the same way as in a TEM.

The photoemission electron microscope has great promise in biology as a high-resolution method for pinpointing the position of different types of molecules in an organic structure. Although this can be done by conventional ultraviolet fluorescence microscopy, the photoemission electron microscope gives a tenfold increase in resolution, to 10–20 nanometres. This does not approach the capability of the TEM, of course, but the photoemission electron microscope has the advantage that it can be used to examine whole cells rather than thin sections of cells.

The first pictures of biological materials produced by the instrument were made in 1972. With advances in the technology of using monoclonal antibodies to label particular molecules, it seems probable that the microscope will gain wider acceptance in the future.

X-ray Microscopy

X-rays have very short wavelengths indeed – in the range from 0.1 nanometres ('hard' X-rays) to 10 nanometres ('soft' X-rays) – but they have two properties that make it difficult to design an X-ray microscope. First, the refractive index of all materials to X-rays is close to unity. This means that no matter how curved a lens-shaped object may be, it will not focus X-rays in a reasonable distance. Second, X-rays carry no electric charge, and it is therefore not possible to focus them using magnetic or electrostatic lenses. These difficulties have been overcome in three different ways.

The simplest is called contact microradiography and it was invented in 1896. The object is placed in contact with a photographic plate, and the whole exposed to X-rays. The result is a life-size image of the object on the photographic plate, which then becomes the 'specimen' viewed in a light microscope. The best techniques of this sort produce a resolution of about 0.5 micrometres.

In point-projection microscopy, the object is placed close to a point source of X-rays produced by irradiating a small piece of a 'target' material such as copper or aluminium. The X-rays radiate in straight lines from the point source, pass through the object, and at some distance away, expose a sheet of photographic film. Magnification depends on the relative distances between the source, the object and the film. All levels in the object are in focus, and a typical application of this method is the study of the vascular system in a tissue or organ. Contrast may be enhanced by the injection of heavy metal salts into the blood stream.

Interest in X-ray microscopy has been renewed by two technical advances. The first is the availability of very intense X-ray beams, obtained from synchrotrons or by laser irradiation of target materials. The intensity of pulsed laser sources is such that it is possible to record an image in a time of only 100 picoseconds (10^{-12} seconds). High-intensity sources also mean that the photographic film or plate can be replaced by a plastic film (called a 'resist'), which is grainless. The image formed within this film can be examined in a TEM. This improves the resolution of contact microradiography to around 100 nanometres.

The second development is the production of Fresnel zone plates for focusing X-rays. Fresnel lenses are familiar optical components, used for example in the viewing screens of many cameras, and as rear-view devices on public transport vehicles. They work by the diffraction of radiation as it passes through a series of finely spaced lines. For X-rays, these lines are only 100 nanometres

The scanning X-ray microscope at the Brookhaven National Laboratory, Long Island. User convenience is a low priority in the design of prototype microscopes.

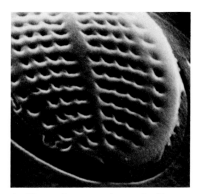

An X-ray microscope picture of a diatom. Even with the most sophisticated technology, diatoms retain their favoured place as test objects for microscopy. The X-ray image is realistically three-dimensional, but in current versions of the microscope the resolution is only about that of an average light microscope.

apart, and the zone plates have to be fabricated using an electron microscope.

The microscope based on their use is a scanning instrument. The zone plate focuses an intense X-ray beam to a point, and the specimen is scanned past this point in a raster pattern, just as in the acoustic microscope. The transmitted X-rays strike a detector, and the output from this is fed to a television screen. Resolution is limited by the accuracy of the movement of the specimen, and results so far indicate that figures in the range 50–200 nanometres are possible.

Microscopes relying on high-intensity sources will never be widely available because of the technology involved in the source itself. Within this severe limitation, however, X-ray microscopy has a potentially fascinating future. In particular, the ability to record an image in a very short space of time means that it should be possible to study dynamic events in biology, such as the condensation of chromosomes and their movement during cell division. The penetrating power of X-rays means that internal defects in crystals can be studied, and processes such as the setting of adhesives.

Scanning Tunnelling Microscopy

A great deal of scientific research centres on the nature of surfaces. The most familiar example is in the semiconductor industry. Microchips consist of very thin films laid one over the other. In such films it is surface properties, not bulk properties, which are of predominant importance.

Similarly, the properties of biological membranes depend on surface interactions between molecules, not gross composition. The SEM produces images of surfaces, but its resolution is limited. This limitation is overcome by the scanning tunnelling microscope.

Its principle is founded in quantum mechanics. Close to the surface of any material there is an increased probability that electrons from surface atoms will travel outside their usual orbits. The surface of a material can be regarded as being covered with a thin cloud of electrons, the thickness and properties of which depend on the arrangement of the surface atoms. The scanning tunnelling microscope investigates this cloud.

A fine-tipped needle (which has its own electron cloud) is brought close to the surface under study. If a voltage is applied between the needle and the object surface, then at a certain point in the approach an electric current will pass across the gap. This is the so-called tunnelling current. Its magnitude is extremely sensitive to the width of the gap; if the distance is increased by the diameter of only one atom, the tunnelling current is diminished by a factor of 1000.

The needle is fitted with a feedback mechanism which controls its vertical motion so as to keep the current constant. If the current falls, the

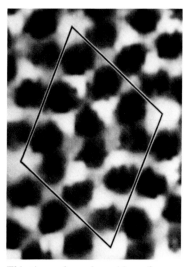

This picture shows the positions of atoms in a crystal of silicon, photographed by a scanning tunnelling microscope. Each black particle in the picture represents a single atom of silicon; the indistinct grey regions between adjacent atoms show the positions of electron bonds which hold the crystal structure together. The magnification is ×90 million.

needle approaches the surface; if it rises, the needle is moved away. The vertical motion is produced by changing the voltage across a piezoelectric crystal, and it is the variations in this voltage which are used to produce the image on the screen of the microscope. Similar crystals provide the horizontal scanning movements across the specimen surface.

The image appears as a three-dimensional line drawing on a computer screen. In the vertical direction, its resolution can be as high as 0.01 nanometres. The sideways resolution is limited by the size of the needle tip; under ideal conditions it can be as high as 0.2 nanometres. Owing to the very small size of the area scanned by the needle, the magnification of the image can be up to 100 million times.

The scanning tunnelling microscope is at present purely a research tool. It has been used to study the surfaces of such materials as silicon, and to produce pictures of gas adsorption onto metals at atomic resolution. The needle and its specimen can be coupled together by means of a layer of very pure water, and so there is no intrinsic difficulty in studying hydrated biological specimens.

Microscopy of the Future

This survey of novel types of microscope is by no means exhaustive. The characteristic they share is specialisation, and many other examples could have been cited. For instance, semiconductor materials are transparent to infrared radiation, and so, inevitably, an infrared microscope has also been designed. The question is, will the laboratories and factories of the future be filled with sophisticated instruments of this type, or will matters remain much as they are now, with the light microscope in its various forms, the TEM and the SEM predominant?

In discussing this question, it is important to remember that microscopy is not merely about microscopes; it is about specimen techniques, and, increasingly, it is about the extraction of information from the microscope image. It is generally true to say that the practical limits of a given microscope at a given time always lie with the specimen rather than the instrument. The exception is during the early development of a new design.

Microscopy is therefore always in the state where existing instruments are more than adequate for the great majority of purposes. Very few practising microscopists spend their days bemoaning the inadequacies of their microscope; rather they are constantly attempting to devise better methods of preparing their specimens.

In the mainstream of microscopy, then, innovation tends to come directly from large manufacturers, either through competitive zeal (a long-standing tradition in microscopy), or through their attempts to increase profitability or market share. Such innovations, paradoxically, tend to be conservative in their outlook: an increase in numerical aperture here, an improvement in flatness of field there. Revolutionary changes tend to be instigated by individual scientists or groups who perceive a theoretical possibility and set about turning it into a working design. After perhaps years, a new type of microscope may emerge. If it has wide enough applications, it may be adopted by commercial manufacturers.

When it comes to the extraction of information from the microscope image, the availability of large computers with frame-store capabilities undoubtedly means that there will be increasing use of computer technology in microscopy. The computer's ability to modify images is immense, provided they are presented to it in an appropriate fashion. With scanning microscopes, the output can be fed directly into the computer, but in instruments with a continuously visible image such as light microscopes and the TEM, a video camera has to be used to relay the image in a linear form.

Programs at present in use or under development include the averaging of successive frames of an image, resulting in a substantial improvement in signal-to-noise ratios; and the production of artificially coloured images to enhance the visual effect of small changes in contrast. The first of these is of great significance, since the ability to control the amount of noise in an image means in practice that the effective sensitivity of a microscope can be increased almost at will. In biological fluorescence microscopy, for example, it should become possible to detect the extremely weak fluorescence due to the presence of rare molecular species.

PICTURE CREDITS

1.1–1.4 Tony Brain/SPL
1.5 David Leah/SPL
1.6 Biophoto Associates
1.7–1.8 Jeremy Burgess/SPL
1.9 Manfred Kage/SPL

2.1 Manfred Kage/SPL
2.2 From *TISSUES AND ORGANS: A Text-Atlas of Scanning Electron Microscopy* by Richard G. Kessel & Randy H. Kardon. Copyright © 1979 W.H. Freeman & Company. Used by permission
2.3 Manfred Kage/SPL
2.4 Biophoto Associates
2.5 Tony Brain/SPL
2.6 G. Schatten/SPL
2.7 David Scharf/SPL
2.8 From *TISSUES AND ORGANS: A Text-Atlas of Scanning Electron Microscopy* by Richard G. Kessel & Randy H. Kardon. Copyright © 1979 W.H. Freeman & Company. Used by permission
2.9–2.10 CNRI/SPL
2.11 Manfred Kage/SPL
2.12–2.13 G. Bredberg/SPL
2.14 Sinclair Stammers/SPL
2.15 M.I. Walker/Science Source/SPL
2.16 Omikron/Science Source/SPL
2.17 Biophoto Associates
2.18 Manfred Kage/SPL
2.19 Biophoto Associates
2.20 Manfred Kage/SPL
2.21 D. Jacobowitz/SPL
2.22 Manfred Kage/SPL
2.23 CNRI/SPL
2.24 Eric Gravé/SPL
2.25 Don Fawcett/Science Source/SPL
2.26 Michael Abbey/Science Source/SPL
2.27 M.I. Walker/Science Source/SPL
2.28 Don Fawcett/Science Source/SPL
2.29 Kevin Fitzpatrick, Guy's Hospital Medical School/SPL
2.30 Eric Gravé/SPL
2.31 Biophoto Associates
2.32 CNRI/SPL
2.33 Tony Brain/SPL
2.34 CNRI/SPL
2.35 Biophoto Associates
2.36 Eric Gravé/SPL
2.37 J. Gennaro/Science Source/SPL
2.38 CNRI/SPL
2.39 J. James/SPL
2.40 Eric Gravé/SPL
2.41 Tony Brain/SPL
2.42 ASA Thorensen/Science Source/SPL
2.43 CNRI/SPL
2.44–2.45 From *CORPUSCLES: Atlas of Red Blood Cell Shapes* by Marcel Bessis, Springer–Verlag, 1974
2.46 A.R. Lawton/SPL
2.47 W. Villiger, Biozentrum/SPL
2.48 A. Liepins/SPL
2.49–2.50 Jeremy Burgess/SPL
2.51–2.52 Biophoto Associates
2.53–2.54 Jeremy Burgess/SPL

3.1 David Scharf/SPL
3.2–3.3 Eric Gravé/SPL
3.4 Biophoto Associates
3.5 Biology Media//Science Source/SPL
3.6 John Walsh/SPL
3.7–3.10 Biophoto Associates
3.11 Cath Wadforth, University of Hull/SPL
3.12–3.13 Kevin Fitzpatrick, Guy's Hospital Medical School/SPL
3.14–3.16 Sinclair Stammers/SPL
3.17–3.20 John Walsh/SPL
3.21 Jeremy Burgess/SPL
3.22 David Scharf/SPL
3.23 Biophoto Associates
3.24 Cath Wadforth, University of Hull/SPL
3.25 Tony Brain/SPL
3.26 Biophoto Associates
3.27–3.30 David Scharf/SPL
3.31–3.32 Jeremy Burgess/SPL
3.33 John Walsh/SPL
3.34–3.38 Jeremy Burgess/SPL
3.39 David Scharf/SPL
3.40 John Walsh/SPL

4.1–4.3 Jeremy Burgess/SPL
4.4 Patrick Lynch/Science Source/SPL
4.5 Chuck Brown Science Source/SPL
4.6 Jeremy Burgess/SPL
4.7–4.8 James Bell/SPL
4.9 Jeremy Burgess/SPL
4.10 M.I. Walker/Science Source/SPL
4.11–4.12 Jeremy Burgess/SPL
4.13 James Bell/SPL
4.14 David Scharf//SPL
4.15–4.22 Jeremy Burgess/SPL
4.23 M.I. Walker/Science Source/SPL
4.24–4.26 Jeremy Burgess/SPL
4.27–4.28 David Scharf/SPL
4.29 Tony Brain/SPL
4.30 Jeremy Burgess/SPL
4.31 R.E. Litchfield/SPL
4.32–4.33 Jeremy Burgess/SPL
4.34 R.E. Litchfield/SPL
4.35 Jeremy Burgess/SPL
4.36 Gene Cox/SPL
4.37–4.40 Jeremy Burgess/SPL
4.41 Biophoto Associates

5.1 Tektoff-RM, CNRI/SPL
5.2 M. Wurtz, Biozentrum/SPL
5.3 Tektoff-RM, CNRI/SPL
5.4 M. Wurtz, Biozentrum/SPL
5.5 Samuel Dales/SPL
5.6 Luc Montagnier, Institut Pasteur, CNRI/SPL
5.7 B. Heggeler, Biozentrum/SPL
5.8 Lee Simon/SPL
5.9 Biozentrum/SPL
5.10 Tony Brain/SPL
5.11 L. Caro/SPL
5.12 Eric Gravé/SPL
5.13–5.14 CNRI/SPL
5.15 Tony Brain/SPL
5.16 CNRI/SPL
5.17 Gopal Murti/SPL
5.18 Jeremy Burgess/SPL
5.19 John Innes Institute/SPL
5.20 Jeremy Burgess/SPL
5.21 Biophoto Associates
5.22 R.B. Taylor/SPL
5.23 Michael Abbey/Science Source/SPL
5.24–5.25 James Bell/SPL
5.26 Jan Hinsch/SPL
5.27–5.28 Biophoto Associates
5.29 Ann Smith/SPL
5.30 Biophoto Associates
5.31 Jeremy Burgess/SPL
5.32 Tony Brain/SPL
5.33–5.34 Biophoto Associates
5.35 Jeremy Burgess/SPL
5.36 Biophoto Associates
5.37 Jeremy Burgess/SPL

6.1 Jeremy Burgess/SPL
6.2 CNRI/SPL
6.3 Jeremy Burgess/SPL
6.4–6.5 Don Fawcett/Science Source/SPL
6.6 Don Fawcett & D. Phillips/Science Source/SPL
6.7–6.8 Don Fawcett/Science Source/SPL
6.9 Don Fawcett & D. Friend/Science Source/SPL
6.10 Don Fawcett/SPL
6.11 Don Fawcett & T. Kuwabara/Science Source/SPL
6.12 Jeremy Burgess/SPL
6.13 K.R. Miller/SPL
6.14 EM Unit, British Museum (Natural History)
6.15 Jeremy Burgess/SPL
6.16 Gopal Murti/SPL
6.17 P. Dawson, John Innes Institute
6.18 Don Fawcett/Science Source/SPL
6.19 Don Fawcett & D. Phillips/Science Source/SPL
6.20–6.21 Biophoto Associates
6.22 Don Fawcett/Science Source/SPL
6.23 Eric Gravé/SPL
6.24 Keith Porter/SPL
6.25 J. Pickett-Heaps/SPL

7.1 Jeremy Burgess/SPL
7.2 Mitsuo Ohtsuki/SPL
7.3 H. Hashimoto, Osaka University
7.4 Y.P. Lin & J.W. Steed, University of Bristol
7.5 I. Baker
7.6 Mike McNamee, Chloride Silent Power Ltd/SPL
7.7 G. Müller, Struers GmbH
7.8 C. Hammond, The University of Leeds
7.9 John P. Pollinger & Gary L. Messing, Ceramic Science Section, Department of Materials Science, The Pennsylvania State University
7.10 St. John & Logan, *J. Crystal Growth* 46, 1979
7.11–7.12 Elizabeth Leistner
7.13 Mike McNamee, Chloride Silent Power Ltd/SPL
7.14 Courtesy of Dr Riedl, Professor Jeglitsch and Dr Locker
7.15 Mike McNamee, Chloride Silent Power Ltd/SPL
7.16 Sydney Moulds/SPL
7.17–7.18 Jeremy Burgess/SPL
7.19 David Parker/SPL
7.20 Jan Hinsch/SPL
7.21–7.28 Mike McNamee, Chloride Silent Power Ltd/SPL
7.29 Lou Macchi, Poroperm-Geochem Ltd
7.30–7.31 Peter Borman, Poroperm-Geochem Ltd
7.32 Jan Hinsch/SPL
7.33–7.36 Mike McNamee, Chloride Silent Power Ltd/SPL
7.37–7.38 G. Müller, Struers GmbH
7.39 Mike McNamee, Chloride Silent Power Ltd/SPL
7.40 G. Müller, Struers GmbH
7.41 Manfred Kage/SPL
7.42–7.43 G. Müller, Struers GmbH

8.1 David Scharf/SPL
8.2–8.5 G. Müller, Struers GmbH
8.6 The Welding Institute, Cambridge Instruments
8.7–8.8 G. Müller, Struers GmbH
8.9 J.D. Williams, Queens University of Belfast
8.10 Max-Planck-Institut fur Metallforschung
8.11 G. Müller, Struers GmbH
8.12 Elizabeth Leistner
8.13 G. Müller, Struers GmbH
8.14 Elizabeth Weidmann, Struers Inc.
8.15 Courtesy of G.N. Babini, A. Bellosi, P. Vincenzini, *J. Mat. Sci.*, 19, 3, 1984
8.16 National Physical Laboratory, Crown Copyright Reserved
8.17 G. Müller, Struers GmbH
8.18 National Physical Laboratory, Crown Copyright Reserved
8.19 G. Müller, Struers GmbH
8.20 David Parker/SPL
8.21 STC/A. Sternberg/SPL
8.22–8.24 Jeremy Burgess/SPL
8.25 VG Semicon/SPL
8.26 Mike McNamee, Chloride Silent Power Ltd/SPL
8.27 Jan Hinsch/SPL
8.28 G. Müller, Struers GmbH
8.29–8.30 Jeremy Burgess/SPL
8.31–8.35 G. Müller, Struers GmbH
8.36 C.E. Price, Oklahoma State University
8.37 J.G. Ashurst, Chloride Silent Power Ltd
8.38 Shell, Thornton Research Centre
8.39 Manfred Kage/SPL
8.40 Courtesy of G.N. Babini, A. Bellosi, P. Vincenzini, *J. Mat. Sci.*, 19, 3, 1984

9.1 Jeremy Burgess/SPL
9.2 R.E. Litchfield/SPL
9.3 Harold Rose/SPL
9.4–9.5 Jeremy Burgess/SPL
9.6 Manfred Kage/SPL
9.7–9.21 Jeremy Burgess/SPL

Page 186 (all pictures) SPL
P. 187 (far & centre left) SPL
P. 187 (centre right) Neil Hyslop
P. 187 (far right) Museum of the History of Science, University of Oxford
P. 188 (left) SPL
P. 188 (centre right) Museum of the History of Science, University of Oxford
P. 188 (far right) GECO UK Ltd/SPL
P. 189 (far left & upper centre) Courtesy of E. Ruska, with thanks to T. Mulvey
P. 189 (lower centre) D. McMullan/SPL
P. 190 (left) D. McMullan/SPL
P. 190 (right) Neil Hyslop
P. 191 Neil Hyslop
P. 192 (upper left) S. Stammers/SPL
P. 192 (upper right, lower left & right) J. Patterson/SPL
P. 193 (upper & lower left) Neil Hyslop
P. 193 (far right) Jan Hinsch/SPL
P. 194 James Stevenson/SPL
P. 195 (upper left) John Durham/SPL
P. 195 (lower left) R. King/SPL
P. 196 Neil Hyslop
P. 197 (left & right) Neil Hyslop
P. 198 (upper left) Heather Davies/SPL
P. 198 (centre right) Muriel Lipman/SPL
P. 198 (far right) Neil Hyslop
P. 199 (far left) Kenneth R. Miller/SPL
P. 199 (upper & lower right) Jeremy Burgess/SPL
P. 200 Norman Costa & Sinclair Stammers/SPL
P. 201 J.G. White, W.B. Amos & M. Fordham (1987) An evaluation of confocal versus conventional imaging of biological structures of fluorescence light microscopy, *J. Cell Biol. (in press)*
P. 202 (left) Lawrence Berkeley Laboratory/SPL
P. 202 (right) Mitsuo Ohtsuki/SPL
P. 203 (right) Neil Hyslop
P. 203 (left) David Parker/SPL
P. 204 (both) Courtesy of IBM

INDEX